Math Concepts
for
Food Engineering
Second Edition

T0186661

Math Concepts
for
Food Engineering
Second Edition

Richard W. Hartel
Robin K. Connelly
Terry A. Howell, Jr.
Douglas B. Hyslop

CRC Press
Taylor & Francis Group
Boca Raton London New York

CRC Press is an imprint of the
Taylor & Francis Group, an **informa** business

CRC Press
Taylor & Francis Group
6000 Broken Sound Parkway NW, Suite 300
Boca Raton, FL 33487-2742

© 2008 by Taylor & Francis Group, LLC
CRC Press is an imprint of Taylor & Francis Group, an Informa business

Library of Congress Cataloging-in-Publication Data

Math concepts for food engineering. -- 2nd ed. / Richard W. Hartel ... [et al.].
    p. cm.
  Includes bibliographical references and index.
  ISBN 978-1-4200-5505-4 (hardback : alk. paper)
  1. Mathematics. 2. Food industry and trade--Mathematics. I. Hartel, Richard W., 1951-

QA37.3.M35 2008
664.001'51--dc22   2007048963

Visit the Taylor & Francis Web site at
http://www.taylorandfrancis.com

and the CRC Press Web site at
http://www.crcpress.com

Contents

Preface to Second Edition

For over 10 years now, we have been assessing the math skills of students entering the first food engineering course in our curriculum. There is a clear correlation between these skills and the student's ability to do well in this course. Students who enter with good math skills generally do well in engineering class. However, those students who need extra help with math can make up the gap through hard work and practice. This is the advantage that *Math Concepts for Food Engineering* brings to those students willing to work at improving their math skills.

For this second edition, we have incorporated some simple food engineering principles within the text. Without going into the detail of a food engineering textbook, some of the more important technical principles have been included relative to the learning outcomes for our food engineering class. We feel that this will give our students a better perspective of the importance of the math skills, and help them better relate these simple problems with the principles they are learning in class.

In this second edition, we have also made several other additions. First, we have incorporated various exercises throughout the text that use spreadsheets, a valuable tool for analyzing and manipulating data. The use of spreadsheets to create mathematical tools of practical use for some applications is developed in chapters 1 through 5, and these are used to help solve some of the examples in chapters 6 through 11, the second part of the text. The publisher will make the spreadsheet exercises seen in the book available on its Web site for those who purchase the book. We have also included a chapter on mass transfer, and added a simple units conversion page in the appendix. This offers a more complete reference for our students by providing complete coverage of basic balance and transport principles used in food engineering.

Math Concepts for Food Engineering, second edition, is still intended as a supplemental reference to a standard textbook for a food engineering class. Its purpose is to provide practice and experience in solving simple engineering problems so that students are better prepared to face the more rigorous problems presented in class.

Richard W. Hartel
Robin K. Connelly
Terry A. Howell, Jr.
Douglas B. Hyslop

The Authors

Richard W. Hartel, PhD, currently a professor of food engineering, has been with the Department of Food Science at the University of Wisconsin–Madison since 1986. Dr. Hartel conducts a research program focused on crystallization and phase transitions in foods, including studies on ice crystallization, sugar crystallization and glass transition, and lipid-phase behavior and crystallization. He teaches courses in food processing, phase transitions, and candy science.

Robin Kay Connelly, PhD, is currently an assistant professor of food engineering with dual appointments with the Departments of Food Science and Biological Systems Engineering at the University of Wisconsin–Madison since 2003. Dr. Connelly conducts research in the areas of computational fluid dynamics, simulation of food processing operations, and food rheology, with an emphasis on mixing applications. She teaches the introductory food engineering principles course for food scientists and biological concepts for engineers.

Terry A. Howell, Jr., received his BS and MS from Texas A&M University in agricultural engineering, with a PhD from the University of Wisconsin–Madison in food engineering. He served on the University of Arkansas Food Science faculty for three years, and now works for McKee Foods Corp. (Collegedale, TN) as manager of research and new technology.

Douglas B. Hyslop received his PhD in food chemistry in 1978 from the University of Wisconsin–Madison with an emphasis on physical surface and colloid chemistry. Since joining the staff at Madison in 1980, he has taught numerous food science courses, including food analysis, food chemistry, food physical chemistry, food colloid chemistry, food processing, and food engineering.

Introduction

Mathematical reasoning is an important skill for a food scientist, but it is a skill that is not shared by all students at the same level. Although all food science students are required to have a comprehensive course in calculus (both integral and differential), often, for one reason or another, mathematical reasoning skills have been lost or forgotten by the time a student reaches the food engineering course.

In the spirit of the Institute of Food Technologists education standards, it makes sense to assess a student's mathematical reasoning skills prior to starting the food engineering course. The intention of this book is to help with that assessment and then to provide assistance for those students who need to brush up on their mathematical skills.

The book is organized into sections that present different materials needed for developing mathematical reasoning skills. The first section (chapters 1–4) covers important mathematical skills needed by students in a food engineering class. These principles are primarily a review of previous math classes (algebra, calculus, etc.). This preliminary section is followed by an important chapter on problem solving (chapter 5). The remaining chapters (6–11) cover food engineering topics likely to be found in a food engineering course for food scientists.

In the back of the book is a short quiz, the screening test (appendix 3a), which can be used to assess mathematical reasoning skills prior to the start of a food engineering course. Students are asked to read through the introductory math sections to refresh their memory on the important concepts prior to taking the screening test. For those students who score low on the screening test or simply want to improve their quantitative skills, the problems provided in *Math Concepts for Food Engineering* are intended to build mathematical confidence, as well as to bridge the simplest math concepts and the more complex engineering principles. As the semester unfolds, students should progress through chapters 6–11 as each topic is covered in class. Students should study each worked problem to make sure they understand the mathematical (and engineering) principles being demonstrated, and then work independently on the accompanying practice problems. To help students apply the principles in *Math Concepts for Food Engineering* to food engineering class, this second edition has included a brief coverage of engineering principles important for the sample problems as a supplement to the main *Food Engineering* textbook.

The following approach is recommended for working through this book:

- Review chapters 1–4 to refresh your knowledge of some important math skills.
- Take the screening test at the back of the book in appendix 3a.
- Grade yourself with the answer key in appendix 3b (after you have completed the entire exam).
- Assess which mathematical skill areas you struggled with.
- Review appropriate materials in chapters 1–4 based on your skills assessment.
- Carefully read through the problem-solving approach in chapter 5. Consider your own approach to mathematical problem solving to see where you might find ways to improve. For example, did you generally follow the steps outlined in chapter 5 when you took the screening test? If not, what might you do differently?
- Work your way through chapters 6 to 11 as each section is covered in a food engineering class. We recommend that you work through the appropriate chapter during the first day or two as each topic is covered in class. In that way, you will be well prepared to solve the more complex problems required in class. If you get stuck on a practice problem, be sure to meet with your instructor for assistance. Although the answers for these practice problems are given in the back of the book, seeking help from your instructor can help you see where you are getting stuck.

In our experience, those students who score low on the screening test but work hard at mastering the mathematical principles covered in *Math Concepts for Food Engineering* go on to do well in food engineering classes. As with most things, the likelihood of success is enhanced when the student is willing to put lots of hard work into learning the material. We hope that *Math Concepts for Food Engineering* provides a resource for you to improve your mathematical reasoning skills and, thereby, to attain greater success in learning food engineering principles.

chapter one

Algebra

Perhaps the easiest way to begin a discussion about algebraic equations and their components is to use examples to point out the various terms. The following expressions will be used to demonstrate the different parts of an equation:

$$y = 3x - 7 \tag{1.1}$$

$$y = 2ax + 3b \tag{1.2}$$

An equation is a mathematical statement that can be read like a sentence. Equation (1.1) may be read as "y equals 3 times x minus 7."

Since both sides of the equation are equivalent, one must always be very careful that any arithmetic operation performed on one side of the equation be performed on the other side as well. This will be stressed in later portions of this chapter.

Equations consist of variables, constants, and arithmetic operators. Mathematically, an equation relates different variables and constants; however, equations become more vital as one understands how physical parameters are linked together through them.

1.1   Variables and constants

1.1.1   Variables

Variables are so named because they are allowed to "vary," and their value may assume different amounts at different times or situations. When one substitutes $x = 1$ into equation (1.1), y is calculated to be −4; however, when $x = 3$ is plugged into equation (1.1), one calculates $y = 2$. One can see that x and y are variables in this equation. Variables may be further categorized into independent and dependent variables. One might be able to logically determine that the value for y in equations (1.1) and (1.2) is *dependent* on the value of x in the equations. That is, as x is *independently* varied, different values for y will be produced. In many situations, the manner in which the equation is written will dictate which variable is dependent and which is independent.

Usually, a term written by itself on the left side of the equation is the dependent variable, while variables on the right side of the equal sign are independent variables. Equation (1.1) can be rewritten by adding 7 to both sides and then dividing both sides by 3 to produce equation (1.3).

$$x = (1/3)y + (7/3) \tag{1.3}$$

Here, y becomes the independent variable and x the dependent variable.

Time is almost always an independent variable. Most biological processes or reactions depend on time. One should be able to recognize that time is the independent variable in virtually every equation in which it appears.

1.1.2 Constants

As the name implies, the values assigned to constants do not change. In equation (1.2) above, "a" and "b" are constants. Some common constants that scientists and engineers encounter include:

> g (gravity): The acceleration due to gravity can almost always be considered constant. Its value in SI units is 9.8 m/s².
>
> N_A (Avogadro's number): This number, representing the number of atoms per mole of a substance, is constant at 6.02×10^{23} atoms/mole.
>
> π (pi): A number used in equations for area, surface area, and volume of rounded objects; also used to convert from angular degrees to radians; approximately equal to 3.14159.

1.2 Equations

1.2.1 Functions

Functions are used to describe a mathematical relationship between an independent variable and one or more constants. They are often represented as f (independent variable). Equation (1.2) may be rewritten as

$$y = f(x), \text{ where } f(x) = 2ax + 3b. \tag{1.4}$$

Equation (1.4) is read as, "y is a function of x, where the function of x is $2ax$ plus $3b$." In a similar fashion, x is a function of y in equation (1.3). In many cases, shorthand notation showing that a variable is a function of another variable is used. Equation (1.4) may also be written as

$$y(x) = 2ax + 3b. \tag{1.5}$$

This form is read identically to that of equation (1.4).

Example 1.2.1

$$y = 2x - 1 \tag{1.6}$$

Which are variables?	x, y
What is the independent variable?	x
What is the dependent variable?	y
Is x a function of a variable as this equation is written?	**no**
Is y a function of a variable as this equation is written?	**yes**

Example 1.2.2

$$V(t) = (g/2)*t + V_0 \tag{1.7}$$

This equation relates the velocity at time, t, to the initial velocity (V_0), acceleration due to gravity (g), and time (t).

Which are constants?	**g, V_0**
Which are variables?	**t, V(t)**
What is the independent variable?	**t**
What is the dependent variable?	**V(t)**
Is V a function of any variable?	**Yes, t**

1.2.2   Manipulation of equations

As stated previously, an equation is a **true** statement, and both sides of the equation must be manipulated in the same way to maintain the validity of the equation. Once one understands the rules and procedures in rearranging equations, one can solve many food engineering problems. The following is an overly simplified example:

$$1 = 1 \tag{1.8}$$

If we want to multiply the left side by 3, we must do it to the right side also so that the statement is still true.

$$3*1 = 3*1$$

$$3 = 3 \tag{1.9}$$

Now, if we wanted to subtract 1 from the left side, we must perform the same operation on the right side of the equation.

$$3 - 1 = 3 - 1$$

$$2 = 2 \tag{1.10}$$

Now, let us move on to the manipulation of a more relevant equation to the food industry, the ideal gas law equation. One of its many forms is as follows:

$$PV = nRT \qquad (1.11)$$

where

> P = pressure (Pa = N/m^2)
> V = volume (m^3)
> n = moles of gas (g·mol)
> R = gas law constant (8.314 N·m/K·mol)
> T = temperature (K)

Which are the variables and constants in this equation? P, V, n, and T are variables, and R is a constant. This equation is very useful, but many times, one would like to isolate one variable from the rest. How would one go about rearranging this equation so that the temperature could be solved, knowing the other four quantities?

To isolate T, we must remove the n and R from the right side of the equation. If we start with n, we must divide the right side by n to remove it and, therefore, must also divide the left side by n. Our equation now looks like:

$$PV*(1/n) = nRT*(1/n)$$

$$PV/n = RT \qquad (1.12)$$

The same procedure must be repeated with R so that the equation takes the final form as shown.

$$(PV/n)*(1/R) = RT*(1/R)$$

$$PV/nR = T \qquad (1.13)$$

If one were given the values for P, V, and n, and knowing the value for the ideal gas law constant, the value for T can now be calculated.

Example 1.2.3

Calculate the temperature for an ideal gas, given:

> P = 200 Pa
> n = 2 moles
> V = 30 m^3

From equation (1.13), $T = PV/nR$
  Substitute known values for P, V, n, and R.

> T = (200 Pa)(30 m^3)/(2 moles)(8.314 N·m/K mol)
> T = 360.83 K

Example 1.2.4

Rearrange the following equation to solve for x. If $y = 4$, find x.

$$y = x^2 - 5$$

First, add 5 to both sides.

$$y + 5 = x^2$$

Now, take the square root of both sides.

$$(y + 5)^{1/2} = x$$

Now solve for $y = 4$.

$$(4 + 5)^{1/2} = x$$
$$(9)^{1/2} = x$$
$$3 = x$$
$$x = 3$$

Example 1.2.5

Fourier's equation is used to determine heat transfer by conduction through a material. A simplified form looks like the following:

$$Q = k*\Delta T/\Delta x \qquad\qquad (1.14)$$

where

   Q = heat transferred per surface area (W/m^2)
   k = thermal conductivity of the material (W/m·K)
   ΔT = change in T across material (K)
   Δx = thickness of material (m)

You are given the heat transferred through a wall, $Q = 1000$ W/m^2; the change in temperature, $\Delta T = 20$ K; and thermal conductivity, $k = 50$ W/m·K. Calculate the thickness of wall required to produce this amount of heat transfer.
   Multiply both sides by Δx.

$$Q*\Delta x = k*\Delta T$$

Divide both sides by Q:

$$\Delta x = k*\Delta T/Q$$

Substitute known values for Q, k, and ΔT:

$$\Delta x = (50 \text{ W/m·K} * 20 \text{ K})/(1000 \text{ W/m}^2) = (1000 \text{ W/m})/(1000 \text{ W/m}^2) = 1 \text{ m}$$

1.2.3 *Rules of equations applied to engineering units*

At this point in the text, it is appropriate to apply this knowledge of equations directly to food engineering concepts. As in all engineering disciplines, problems are solved by using equations that apply to the unique situation at hand. In the case of food engineering, the equations are typically related to mass, heat, and energy balances and transfer. The equations that govern these phenomena are covered in this and other texts. As has been seen in the previous examples of equations, food engineering equations contain constants and variables, both dependent and independent.

The concept of units sets apart engineering and scientific equations from basic algebra. Whereas algebra looks at equations and mathematics in a pure sense, science and engineering apply the concepts to real problems. Units are vital to add understanding to the results that come from solving an engineering problem. In fact, most engineering solutions are invalid without providing the units of measure for the answer. If a colleague asks you for the flow rate of corn syrup to a mixing vessel, there is a huge difference between 10 (meaningless without units), 10 liters/hour, and 10 gallons/hour.

When dealing with units, the food engineer must be careful to use a common system while solving problems. There are two main systems, and the reader most likely has some familiarity with both the English system (still used predominantly in the United States), and the Système International (SI or metric unit system, used more commonly worldwide). Usually, a problem and its solution is given in one of these unit systems, but occasionally, a problem may be presented with some variable(s) in SI and other(s) in English units. In addition, even within one of the systems, several different units can be used to describe the same measure (e.g., inches, feet, yards, etc., are all English units for length). The remainder of this section will detail how to easily convert from one system to another and within one system.

Starting with a simple example, recall that 1 ft is equivalent to 12 in. Writing this in an equation form yields

$$1 \text{ ft} = 12 \text{ in.} \tag{1.15}$$

By definition, the equation must be true, and it can be manipulated like any other. When a manipulation is performed, the units must be carried with the manipulation. So, to generate a relationship that shows how many inches are in one foot, equation (1.15) can be manipulated by dividing both sides by 1 ft.

$$(1 \text{ ft}/1 \text{ ft}) = 1 = (12 \text{ in}/1 \text{ ft}) \tag{1.16}$$

The left side of equation (1.16) reduces to 1 (with no units) because the units of feet are divided by units of feet (this is typically called "canceling units"). Thus, when unit relationships like that above are developed, they can be used as if multiplying by 1. Now, the relationship on the right side of equation (1.16) may be used to convert any length in feet to inches by simply multiplying with it and canceling units. Again, it is as if 1 is being multiplied by the length in feet.

Example 1.2.6

1. Use equation (1.16) to convert 5 ft to units of inches.
2. Also, rearrange equation (1.15) to develop a relationship to convert from inches to feet, and then convert 42 in. to feet.
    1. Convert feet to inches.

$$5 \text{ ft} * (12 \text{ in.}/1 \text{ ft}) = \textbf{60 in.}$$

   2. Dividing both sides of equation (1.15) by 12 in. gives

$$(1 \text{ ft}/12 \text{ in.}) = (12 \text{ in.}/12 \text{ in.}) = 1 \tag{1.17}$$

$$42 \text{ in.} * (1 \text{ ft}/12 \text{ in.}) = \textbf{3.5 ft.}$$

These are obviously simple examples, but they show how the concept works. The concept can be applied to any units conversion that is required.

This example shows the technique to use when converting between primary units — those measuring the most basic of dimensions. Primary units describe things like length, mass, time, and temperature. Secondary units (or derived units) are used to describe quantities that are combinations of primary units. The units of velocity (distance divided by time) comprise a secondary unit, regardless of the actual primary units used to express it. When a scientist needs to convert secondary units from one system to another, one can use the method above for each primary unit to convert the derived unit. Appendix A contains a brief listing of some useful unit conversions in food engineering. Also, there are some very complete conversion tables available that will also show the direct relationship between secondary units in different systems (e.g., the units of energy, 1 Btu/s = 1.0545 kW).

Example 1.2.7

The density of soybean oil is 920 kg/m^3. Convert this to English units (lb$_m$/ft^3). Use only primary units in one solution. Use a secondary-to-secondary conversion in another solution.

$$1 \text{ m} = 3.2808 \text{ ft}$$

$$1 \text{ lb}_m = 0.45359 \text{ kg}$$

$$920 \text{ kg/m}^3 * (1 \text{ m})^3/(3.2808 \text{ ft})^3 = 26.05 \text{ kg/ft}^3$$

$$26.05 \text{ kg/ft}^3 * (1 \text{ lb}_\text{m}/0.45359 \text{ kg}) = \textbf{57.44 lb}_\textbf{m}\textbf{/ft}^\textbf{3}$$

Notice that in converting m³ to ft³, the entire quantity had to be cubed to correctly make the conversion. Also, since the m³ term was in the denominator in the original unit, the conversion factor needed the m³ term in the numerator to be canceled properly. Understanding these two concepts will serve the scientist or engineer very well.

$$1 \text{ lb}_\text{m}/\text{ft}^3 = 16.019 \text{ kg/m}^3$$

$$920 \text{ kg/m}^3 * [1 \text{ lb}_\text{m}/\text{ft}^3/(16.019 \text{ kg/m}^3)] = \textbf{57.43 lb}_\textbf{m}\textbf{/ft}^\textbf{3}$$

One last technique that may help erase the confusion that sometimes occurs when converting units involves using a "table" method. The technique allows the user to see the conversions a little easier, thus ensuring that each unit is cancelled properly. It is demonstrated below using the previous example:

$$(920 \text{ kg/m}^3) [(1 \text{ m})^3/(3.2808 \text{ ft})^3] (1 \text{ lb}_\text{m}/0.45359 \text{ kg}) = \textbf{57.44 lb}_\textbf{m}\textbf{/ft}^\textbf{3}$$

Setting up units conversion problems like this allows the scientist to quickly ensure that the units will cancel properly and that everything is in order before actually solving the problem. Once the technique has been practiced, it will become second nature.

1.3   Linear and nonlinear equations

The standard equation of a straight line is

$$y = mx + b \tag{1.18}$$

where

  y = any dependent variable
  x = any independent variable
  m = the slope (constant)
  b = the y-intercept (constant, takes the value of y when x equals zero).

If an equation fits this standard form, it may be considered linear. Note that occasionally the equation must be rearranged to fit this form. If an equation has a different form than this standard equation, it is nonlinear.

Example 1.3.1

$$y = 2x + 2$$

Is this equation linear? If so, find the slope and y-intercept.

The equation has the standard linear equation form. The **slope** is **2** and the **y-intercept** is **2**.

In many cases, a nonlinear equation can be manipulated to produce a pseudo-linear equation by replacing the "nonlinear" variable with a new variable that is linear. Example 1.3.2 illustrates this idea.

Example 1.3.2

$$y = x^2 - 3$$

Is this a linear equation? If not, can you linearize it in another form? Determine the slope and y-intercept of the linear equation.

In this form, the equation is **nonlinear** in x; however, it can be linearized by creating a new variable. Let $u = x^2$. The equation now looks like $y = u - 3$, which has the linear equation format with respect to u. The **slope** is **1** and the **y-intercept** is **−3**.

Example 1.3.3

The following equation is used to model the viscosity of a Herschel–Bulkley fluid, one that exhibits a yield stress and power law characteristics such as toothpaste.

$$\sigma = k\gamma^n + \sigma_0 \qquad (1.19)$$

where

σ = shear stress in the fluid (Pa) = (N/m^2)
k = fluid consistency index (Pa s^n)
n = flow behavior index (no units)
γ = shear rate (1/s)
σ_0 = yield stress (Pa).

What are the constants and variables in this equation? Is this a linear equation in σ? If not, can you linearize it?

The constants are k, n, and σ_0, while the shear stress and shear rate are variables.

Since the form of the equation does not fit the standard linear form, the equation is not linear. One method for linearizing this equation involves substituting $u = \gamma^n$ into the equation. The equation now has the form:

$$\sigma = ku + \sigma_0 \qquad (1.20)$$

where equation (1.20) is linear in σ. Another technique involves taking the logarithm of both sides of equation (1.19), and this concept will be covered in section 3.3.

1.4   Multiple linear equations

A common type of problem encountered in food engineering consists of solving sets of linear equations for the values of the variables common in each equation. For instance, in a mass-balance problem, one might want to solve for the amount of product A to add to product B to get a desired product C. Some very simple problems can be solved intuitively, but as the problems become more difficult, one must be able to confidently use methods of solving linear sets of equations.

The most important concept to understand when solving sets of equations is that one must have a separate equation for each unknown variable. That is, if one wants to solve for x and y, two equations containing x and y are necessary. If one wants to solve for x, y, and z, three separate equations relating x, y, and z are needed.

Three common methods exist for solving sets of equations: algebraic manipulation, substitution, and matrices (Kramer's rule). Additionally, graphing a pair of equations and locating their intersection will produce a correct solution; however, this method is not very common today.

1.4.1   Algebraic manipulation

Recall that equations are true statements. Thus, two equations can be added together without causing any inequality to occur.

$$a = b \tag{1.21}$$

$$c = d \tag{1.22}$$

Adding equation (1.21) and equation (1.22) produces either

$$a + c = b + d \tag{1.23}$$

or

$$a + d = b + c \tag{1.24}$$

depending on how the equations were added.

As you can see, the "equality" of each equation is not compromised as long as equivalent actions are performed on each side. This is the basis of the algebraic manipulation technique in solving sets of equations. The equations are rearranged and added together to solve for one variable. The result obtained from adding or subtracting equations can be substituted for one of the given equations. This is repeated until one variable is easily calculated. We will now work through an example solving two linear equations for two unknown variables. This example is analogous to a set of equations that would be generated when mixing two ingredients with different compositions. The goal is to find the values of x and y (which would correspond to ingredients in this case). If the two equations are independent of each other (that is, if they are not just the same equation written in a different form), there is exactly one value for x and y that solves the equations.

$$2x + y = 5 \tag{i}$$

$$x - y = 1 \tag{ii}$$

Solve for x and y.

The initial step of this method is to get the coefficient in front of one variable in the first equation equal to the negative value of the coefficient of the same variable in the second equation. In this example, the coefficients in front of y fit this criterion already. If (i) is added to (ii), (iii) is produced.

$$3x = 6 \tag{iii}$$

Dividing both sides by 3 gives

$$x = 2. \tag{iv}$$

Substituting 2 for x in (ii) gives $y = 1$. That was easy. Now let us rework the problem by isolating the coefficients in front of x instead of y. Multiply (ii) by -2 to get (v) and replace (ii) with (v).

$$2x + y = 5 \tag{i}$$

$$-2x + 2y = -2 \tag{v}$$

Add (i) and (v) to produce (vi).

$$3y = 3 \tag{vi}$$

Divide both sides by 3 to give (vii).

$$y = 1 \tag{vii}$$

Substituting 1 for y in (ii) produces $x = 2$, the same result as before. The problem has been worked through for both variables to show that it does not matter from which direction one approaches the problem, as long as the proper rules for handling equations are applied.

Now, we are ready to move on to the three-variable problem that might represent mixing of three ingredients with different compositions.

$$2x - y + 2z = 6 \tag{a}$$

$$3x + 2y - z = 16.5 \tag{b}$$

$$-2x + 4y + 4z = 24 \tag{c}$$

Find the values for x, y, and z that satisfy these three equations.

The first step in this problem is to eliminate one variable from two of the equations so that a two-variable problem is left (with two equations). This "new" problem can then be solved as done previously. Adding (a) and (c) will eliminate x from one equation. The equations now look like this after replacing (c) with the result of the addition:

$$2x - y + 2z = 6 \tag{a}$$
$$3x + 2y - z = 16.5 \tag{b}$$
$$3y + 6z = 30 \tag{d}$$

We now want to remove x from another equation. Multiply (a) by 3 and (b) by –2.

$$6x - 3y + 6z = 18 \tag{e}$$
$$-6x - 4y + 2z = -33 \tag{f}$$
$$3y + 6z = 30 \tag{d}$$

Add (e) and (f), and replace (f) with the result.

$$2x - y + 2z = 6 \tag{a}$$
$$-7y + 8z = -15 \tag{g}$$
$$3y + 6z = 30 \tag{d}$$

Solve (d) and (g) for y and z. Multiply (g) by 3 and (d) by 7.

$$-21y + 24z = -45 \tag{h}$$
$$21y + 42z = 210 \tag{i}$$

Add (h) and (i).

$$66z = 165 \tag{j}$$

Divide by 66.

$$z = 2.5$$

Substituting 2.5 for z in (d) gives $y = 5$. Substituting these values for y and z into (a) gives $x = 3$. One can check all three equations with the values for x, y, and z to prove the answer is correct.

The original equations are as follows:

$$2x - y + 2z = 6 \tag{a}$$

$$3x + 2y - z = 16.5 \tag{b}$$

$$-2x + 4y + 4z = 24 \tag{c}$$

Now, after inserting the values for x, y, and z, one can see whether the solution is correct.

$$2(3) - 5 + 2(2.5) = 6 - 5 + 5 = 6 \tag{a}$$

$$3(3) + 2(5) - 2.5 = 9 + 10 - 2.5 = 16.5 \tag{b}$$

$$-2(3) + 4(5) + 4(2.5) = -6 + 20 + 10 = 24 \tag{c}$$

The answer checks.

1.4.2   Substitution

This method is probably the easiest of the three demonstrated here. It consists of rearranging one or more of the equations to isolate one of the unknown variables. This new equation can then be substituted into the remaining equations. This procedure is repeated until the problem is easily solved.

For the same two equations used earlier:

$$2x + y = 5 \tag{i}$$

$$x - y = 1 \tag{ii}$$

Solve for x and y.

We will again solve this problem twice to demonstrate this method. In trying to isolate x or y, the problem becomes easier when x or y in one of the equations has 1 or –1 for its coefficient. In equation (ii), this is satisfied for x and y. We will choose to isolate x by rewriting (ii) as follows:

$$x = 1 + y \tag{iii}$$

We now substitute this equation into (i) to get (iv).

$$2(1 + y) + y = 5 \tag{iv}$$

Rearranging gives

$$2 + 2y + y = 5$$

$$3y + 2 = 5$$

$$3y = 3$$

$$y = 1.$$

Substituting $y = 1$ into (iii) gives $x = 2$. To work this from a different "direction," rewrite (i) to isolate y.

$$y = 5 - 2x \qquad \text{(v)}$$

Substitute this into (ii) to get (vi)

$$x - (5 - 2x) = 1. \qquad \text{(vi)}$$

Now, solve for x.

$$3x - 5 = 1$$

$$3x = 6$$

$$x = 2$$

Substituting $x = 2$ into (v) gives $y = 1$; the same as before.

This method works the same way with three variables. We will use the same set of equations to solve the three-variable problem as used in the algebraic manipulation section 1.4.1.

$$2x - y + 2z = 6 \qquad \text{(a)}$$

$$3x + 2y - z = 16.5 \qquad \text{(b)}$$

$$-2x + 4y + 4z = 24 \qquad \text{(c)}$$

Find the values for x, y, and z that satisfy these three equations.

The coefficient for y in (a) is -1. This makes it a prime target to isolate. From equation (a), an equation for y may be written as

$$y = 2x + 2z - 6. \qquad \text{(d)}$$

Substitute this into (b) and (c).

$$3x + 2(2x + 2z - 6) - z = 16.5 \qquad \text{(e)}$$

$$-2x + 4(2x + 2z - 6) + 4z = 24 \qquad \text{(f)}$$

Reducing these equations,

$$3x + 4x + 4z - 12 - z = 16.5 \qquad \text{(e)}$$

$$-2x + 8x + 8z - 24 + 4z = 24 \qquad \text{(f)}$$

$$7x + 3z = 28.5 \qquad \text{(e)}$$

$$6x + 12z = 48. \qquad \text{(f)}$$

Isolate x from equation (f).

$$6x = 48 - 12z$$

$$x = 8 - 2z \tag{g}$$

Substitute (g) into (e) and solve for z.

$$7(8 - 2z) + 3z = 28.5 \tag{h}$$

$$56 - 14z + 3z = 28.5$$

$$-11z = -27.5$$

$$11z = 27.5$$

$$z = 2.5$$

Substitute $z = 2.5$ into (g) to obtain $x = 3$. Substitute the values for x and y into (d) to obtain

$$y = 5.$$

1.4.3 *Matrices (Kramer's rule)*

Using Kramer's rule to solve for unknowns in a set of equations involves the use of matrices. The set of equations to be solved are converted into a matrix, and some simple rules are used to solve for the unknown variables. This section will go through some of the basics of matrix algebra and definitions of key terms.

A **matrix** is a collection of numbers in an ordered format. A matrix is usually represented by an uppercase letter with its components resembling the following:

For a 2×2 matrix,

$$A = \begin{bmatrix} a_{11} & a_{12} \\ a_{21} & a_{22} \end{bmatrix}. \tag{1.25}$$

For a 3×3 matrix,

$$B = \begin{bmatrix} b_{11} & b_{12} & b_{13} \\ b_{21} & b_{22} & b_{23} \\ b_{31} & b_{32} & b_{33} \end{bmatrix}. \tag{1.26}$$

Matrices are referred to by listing the number of columns and rows. Matrix A above is a 2×2 matrix, while B is a 3×3 matrix. One will notice that the matrices presented so far are "square" matrices. That is, they have the same number of rows as columns. Matrices may exist in any "rectangular" form, where the numbers of rows and columns are different from each other. The only other type of matrix encountered in this section will be one dimensional. These are commonly called arrays or column vectors. Any single element in an array may be referenced by its individual indices (row, column). That is, one may refer to the element in the first row and second column of B above as b_{12}. Each individual element is independent of those around it and functions separately from those around it in matrix algebra and calculations.

1.4.3.1   Addition and subtraction

Matrices may be added together only if they have the same dimensions. A 2×2 matrix cannot be added to a 3×3 matrix, but only may be added to another 2×2 matrix. The addition is performed by adding the elements with the same subscripts from the addend matrices to each other and placing each of these sums in their respective elements in the new matrix. Subtraction is performed in the same manner.

Example 1.4.1

Using the following two matrices, G and H, perform the following algebraic steps:

$$G + H = J$$

$$G - H = K$$

$$H - G = L$$

Solution:

$$G = \begin{bmatrix} g_{11} & g_{12} \\ g_{21} & g_{22} \end{bmatrix} = \begin{bmatrix} 2 & 3 \\ 5 & -2 \end{bmatrix}$$

$$H = \begin{bmatrix} h_{11} & h_{12} \\ h_{21} & h_{22} \end{bmatrix} = \begin{bmatrix} -1 & 3 \\ 4 & -2 \end{bmatrix}$$

$$J = G + H = \begin{bmatrix} (g_{11} + h_{11}) & (g_{12} + h_{12}) \\ (g_{21} + h_{21}) & (g_{22} + h_{22}) \end{bmatrix} = \begin{bmatrix} (2 + (-1)) & (3 + 3) \\ (5 + 4) & ((-2) + (-2)) \end{bmatrix} = \begin{bmatrix} 1 & 6 \\ 9 & -4 \end{bmatrix}$$

$$K = G - H = \begin{bmatrix} (g_{11} - h_{11}) & (g_{12} - h_{12}) \\ (g_{21} - h_{21}) & (g_{22} - h_{22}) \end{bmatrix} = \begin{bmatrix} (2 - (-1)) & (3 - 3) \\ (5 - 4) & ((-2) - (-2)) \end{bmatrix} = \begin{bmatrix} 3 & 0 \\ 1 & 0 \end{bmatrix}$$

$$L = H - G = \begin{bmatrix} (h_{11} - g_{11}) & (h_{12} - g_{12}) \\ (h_{21} - g_{21}) & (h_{22} - g_{22}) \end{bmatrix} = \begin{bmatrix} ((-1) - 2) & (3 - 3) \\ (4 - 5) & ((-2) - (-2)) \end{bmatrix} = \begin{bmatrix} -3 & 0 \\ -1 & 0 \end{bmatrix}$$

1.4.3.2   Multiplication

One may multiply matrices with constants and other matrices. Multiplication of a matrix by a constant is a straightforward process. Each element of the matrix is multiplied by the respective constant, and these individual products are the elements of the new matrix.

Example 1.4.2

Calculate the product of 2*G.

$$2*G = \begin{bmatrix} 2*g_{11} & 2*g_{12} \\ 2*g_{21} & 2*g_{22} \end{bmatrix} = \begin{bmatrix} 2*2 & 2*3 \\ 2*5 & 2*(-2) \end{bmatrix} = \begin{bmatrix} 4 & 6 \\ 10 & -4 \end{bmatrix}$$

Multiplication of a matrix with another matrix follows one very important rule: the number of columns in the first matrix must equal the number of rows in the second matrix. This rule must be followed without exception.

$$C_{mn} D_{np} = E_{mp} \tag{1.27}$$

$$\begin{bmatrix} c_{11} & c_{12} & \cdots & c_{1n} \\ c_{21} & c_{22} & \cdots & c_{2n} \\ \vdots & \vdots & \ddots & \vdots \\ c_{m1} & c_{m2} & \cdots & c_{mn} \end{bmatrix} \begin{bmatrix} d_{11} & d_{12} & \cdots & d_{1p} \\ d_{21} & d_{22} & \cdots & d_{2p} \\ \vdots & \vdots & \ddots & \vdots \\ d_{n1} & d_{n2} & \cdots & d_{np} \end{bmatrix} = \begin{bmatrix} e_{11} & e_{12} & \cdots & e_{1p} \\ e_{21} & e_{22} & \cdots & e_{2p} \\ \vdots & \vdots & \ddots & \vdots \\ e_{m1} & e_{m2} & \cdots & e_{mp} \end{bmatrix}$$

Multiplying a 2×2 matrix with a 2×1 matrix will produce a 2×1 matrix. Likewise, multiplying a 3×3 matrix by a 3×1 matrix produces a 3×1 matrix. On the other hand, a 2×2 matrix cannot be multiplied with a 3×3 matrix or a 3×1 matrix. This multiplication rule also implies that the order of matrices to be multiplied is very important. Two matrices might be compatible in one configuration but not in another. In fact, the only time order is not important is when the two matrices to be multiplied are both square matrices of the same dimension.

Example 1.4.3

W is a 3×3 matrix; X is a 3×2 matrix; Y is a 2×2 matrix; and Z is a 2×1 matrix. Are the following combinations possible?

    WX = yes, a 3×3 matrix times a 3×2 matrix produces a 3×2 matrix
    WY = no, a 3×3 matrix and a 2×2 matrix are incompatible
    WZ = no, a 3×3 matrix and a 2×1 matrix are incompatible
    XW = no, a 3×2 matrix and a 3×3 matrix are incompatible
    XY = yes, a 3×2 matrix times a 2×2 matrix produces a 3×2 matrix

XZ = yes, a 3×2 matrix times a 2×1 matrix produces a 3×1 matrix
YW = no, a 2×2 matrix and a 3×3 matrix are incompatible
YX = no, a 2×2 matrix and a 3×2 matrix are incompatible
YZ = yes, a 2×2 matrix times a 2×1 matrix produces a 2×1 matrix
ZW = no, a 2×1 matrix and a 3×3 matrix are incompatible
ZX = no, a 2×1 matrix and a 3×2 matrix are incompatible
ZY = no, a 2×1 matrix and a 2×2 matrix are incompatible

One can use equation (1.27) to anticipate how many rows and columns will be generated in a matrix multiplication operation. Using the theoretical matrices in equation (1.27), equation (1.28) shows how the individual elements of E are generated.

$$e_{ij} = \sum_{k=1}^{n} c_{ik} d_{kj} = c_{i1} d_{1j} + c_{i2} d_{2j} + \ldots + c_{in} d_{nj}$$

(1.28)

where

$i = 1, \ldots, m$
$j = 1, \ldots, p$

Each individual element of E (e_{ij}) must then be calculated with this formula. For examples and problems in this book, 3×3 matrices will be the largest matrix size encountered, and m and p above will never be larger than three. In words, the individual elements of E (e_{ij}) are found by summing the products of the elements of the ith row of C by the jth column of D. Although this looks messy, matrix multiplication is relatively easy in two dimensions, and one should become proficient in performing these types of operations.

The following example present the steps required to multiply two matrices.

Example 1.4.4

$$A = \begin{bmatrix} 2 & 4 \\ 3 & 3 \end{bmatrix}$$

$$B = \begin{bmatrix} x \\ y \end{bmatrix}$$

Are these matrices compatible for multiplication? If so, what is the product of AB?

A is a 2×2 matrix and B is a 2×1 matrix. They are compatible and will produce a 2×1 matrix:

$$AB = C$$

Following the notation of equation (1.27), $m = 2$, $n = 2$, and $p = 1$. Refer to equation (1.28) for the elements of C.

$$c_{11} = \sum_{k=1}^{2} a_{1k}b_{k1} = a_{11}b_{11} + a_{12}b_{21} = 2x + 4y$$

$$c_{21} = \sum_{k=1}^{2} a_{2k}b_{k1} = a_{21}b_{11} + a_{22}b_{21} = 3x + 3y$$

This produces

$$C = \begin{bmatrix} c_{11} \\ c_{21} \end{bmatrix} = \begin{bmatrix} (2x + 4y) \\ (3x + 3y) \end{bmatrix}$$

a 2×1 matrix with elements defined by the algebraic equations:

$$c_{11} = 2x + 4y$$

$$c_{21} = 3x + 3y$$

Hopefully, by seeing how this matrix is put together, one will be able to see how it is pulled apart later.

The typical set of linear equations in two dimensions resembles the following:

$$ax + by = c$$

$$dx + ey = f$$

This can be written in matrix form as the following:

$$\begin{bmatrix} a & b \\ d & e \end{bmatrix} \begin{bmatrix} x \\ y \end{bmatrix} = \begin{bmatrix} c \\ f \end{bmatrix}$$

Kramer's rule requires calculation of the determinant of the matrix defined by the coefficients (a, b, d, e).

A **determinant** is a function (used with matrices) to convert a matrix into a scalar. Kramer's rule uses determinants to solve sets of equations. It is also used to determine whether a set of equations is linearly independent. It is usually written as "det A," where A is a square matrix. If det $A = 0$, the set of equations is not linearly independent, and infinite solutions exist.

To calculate the determinant for a 2×2 matrix, one uses a simple formula. Using the matrix from equation (1.25),

$$\det A = a_{11} a_{22} - a_{21} a_{12}.$$

The formula for the determinant of a 3×3 matrix is very similar, with a little twist. Using the matrix of equation (1.26),

$$\det B = b_{11} (b_{22} b_{33} - b_{32} b_{23}) - b_{12} (b_{21} b_{33} - b_{31} b_{23}) + b_{13} (b_{21} b_{32} - b_{31} b_{22}).$$

Example 1.4.5

$$H = \begin{bmatrix} 5 & 4 \\ -2 & 3 \end{bmatrix}$$

Calculate the determinant of H.

$$\det H = h_{11} h_{22} - h_{21} h_{12}$$

$$\det H = 5*3 - (-2)*4 = 15 + 8 = 23$$

Example 1.4.6

$$P = \begin{bmatrix} p_{11} & p_{12} & p_{13} \\ p_{21} & p_{22} & p_{23} \\ p_{31} & p_{32} & p_{33} \end{bmatrix} = \begin{bmatrix} 4 & 7 & 6 \\ 3 & -9 & 1 \\ 2 & 3 & -2 \end{bmatrix}$$

Calculate the determinant of P.

$$\det P = p_{11} (p_{22} p_{33} - p_{32} p_{23}) - p_{12} (p_{21} p_{33} - p_{31} p_{23}) + p_{13} (p_{21} p_{32} - p_{31} p_{22})$$

$$\det P = 4*((-9)*(-2) - 3*1) - 7*(3*(-2) - 2*1) + 6*(3*3 - 2*(-9))$$

$$= 4*(18 - 3) - 7*(-6 - 2) + 6*(9 + 18) = 4*(15) - 7*(-8) + 6*(27)$$

$$= 60 + 56 + 162$$

$$= 278$$

To solve a set of linear equations using Kramer's rule, one begins by finding the determinant of the coefficient matrix (the matrix formed by the coefficients of the variables in the equations). To solve for the first unknown variable, one then substitutes the column vector (constant one-dimensional matrix formed by the right-hand side of the linear equations) into the spots for the unknown variable and calculates a new determinant. The value for the variable is then the quotient produced by dividing the new determinant by the determinant of the coefficient matrix. This same procedure is repeated for each unknown variable. The following hypothetical example will demonstrate this method:

$$Ax + By = E$$

$$Cx + Dy = F$$

Here, the coefficient matrix is $\begin{bmatrix} A & B \\ C & D \end{bmatrix}$, and the column vector is $\begin{bmatrix} E \\ F \end{bmatrix}$. The determinant of the coefficient matrix is $AD - CB$. The new matrix created by the substitution of the column vector into the coefficient matrix at the x-coefficient positions is $\begin{bmatrix} E & B \\ F & D \end{bmatrix}$. The determinant of this matrix is $ED - FB$. To calculate x, one takes the quotient of these two matrices:

$$x = (ED - FB)/(AD - CB)$$

The value for y is calculated in the same way as

$$y = (AF - CE)/(AD - CB).$$

Now, let us work through our previous example with two equations.

$$2x + y = 5 \tag{i}$$

$$x - y = 1 \tag{ii}$$

Solve for x and y.

1. Rewrite in matrix form.

$$\begin{bmatrix} 2 & 1 \\ 1 & -1 \end{bmatrix} \begin{bmatrix} x \\ y \end{bmatrix} = \begin{bmatrix} 5 \\ 1 \end{bmatrix}$$

2. Calculate the determinant of the coefficient matrix.

$$\begin{bmatrix} 2 & 1 \\ 1 & -1 \end{bmatrix}$$

det coeff. $= 2*(-1) - 1*1 = -2 - 1 = -3$

3. Substitute column vector, $\begin{bmatrix} 5 \\ 1 \end{bmatrix}$, into "x" positions in the coefficient matrix.

$$\begin{bmatrix} 5 & 1 \\ 1 & -1 \end{bmatrix}$$

4. Calculate the determinant of this matrix.

$$\det = (5* - 1) - (1*1) = -5 - 1 = -6$$

5. Calculate x as the ratio of the determinant of the substituted matrix to the determinant of the coefficient matrix.

$$x = (-6)/(-3) = 2$$

6. Substitute column vector, $\begin{bmatrix} 5 \\ 1 \end{bmatrix}$, into "y" positions in the coefficient matrix.

$$\begin{bmatrix} 2 & 5 \\ 1 & 1 \end{bmatrix}$$

7. Calculate the determinant of this matrix.

$$\det = (2*1) - (1*5) = 2 - 5 = -3$$

8. Calculate y as the ratio of the determinant of the substituted matrix to the determinant of the coefficient matrix.

$$y = (-3)/(-3) = 1$$

Thus,

$$(x, y) = (2, 1)$$

as shown in previous methods.

Solving equations using matrices lends itself well to spreadsheet solutions. Since the steps required to solve the equations are mechanical, macros or templates could be used for common problems. A spreadsheet solution to Kramer's rule can be found at the end of the next example.

Example 1.4.7

A moist food containing 80% water is dried to 50% moisture. The final product weighs 40 lb. How much product entered the drier, and how much water was removed? (See Figure 1.1.)

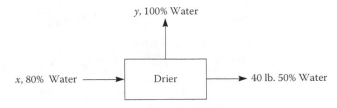

Figure 1.1 Schematic diagram of drier

Since two unknowns need to be calculated, two equations need to be generated. These equations are created by balancing the total amount of material and the amount of water going through the system.

Total mass balance:

$$x = y + 40 \text{ lb}$$

or

$$x - y = 40 \text{ lb} \tag{a}$$

Mass balance on water:

$$0.8\,x - y = 0.5(40 \text{ lb}) = 20 \text{ lb} \tag{b}$$

These two material balances give two equations with two unknown variables.

Now, use the three methods presented (manipulation, substitution, and matrices [Kramer's rule]) to solve the above set of equations for x and y.

1. Manipulation
   Subtract (b) from (a) to create (c)

$$0.2x = 20 \text{ lb} \tag{c}$$

Solve for x.

$x = 20 \text{ lb}/0.2 = 100 \text{ lb}$ (amount of material entering drier)

Substitute x into (a) to solve for y.

$100 - y = 40 \text{ lb}$

$y = 60 \text{ lb}$ (amount of water leaving the drier)

Solution:

$$(x,y) = (100 \text{ lb}, 60 \text{ lb})$$

2. Substitution
   Rewrite (a) to isolate x

$$x = 40 \text{ lb} + y \tag{e}$$

Substitute (e) into (b).

$$0.8(40 \text{ lb} + y) - y = 20 \text{ lb}$$

Expand

$$32 \text{ lb} + 0.8\,y - y = 20 \text{ lb}$$

Solve for *y*.

$$-0.2y = -12 \text{ lb}$$
$$0.2y = 12 \text{ lb}$$
$$y = 12 \text{ lb}/0.2 = 60 \text{ lb}$$

Substitute *y* into (a) to solve for *x*

$$x = 40 \text{ lb} + 60 \text{ lb} = 100 \text{ lb}$$

Solution:

$$(x, y) = (100 \text{ lb}, 60 \text{ lb})$$

3. Matrices (Kramer's rule)
   Rewrite in matrix form.

$$\begin{bmatrix} 1 & -1 \\ 0.8 & -1 \end{bmatrix}\begin{bmatrix} x \\ y \end{bmatrix} = \begin{bmatrix} 40 \text{ lb} \\ 20 \text{ lb} \end{bmatrix}$$

Calculate the determinant of the coefficient matrix.

$$\det = 1*(-1) - (0.8)*(-1) = -0.2$$

Substitute column vector, $\begin{bmatrix} 40 \text{ lb} \\ 20 \text{ lb} \end{bmatrix}$, into "x" positions in the coefficient matrix.

$$\begin{bmatrix} 40 \text{ lb} & -1 \\ 20 \text{ lb} & -1 \end{bmatrix}$$

Calculate the determinant of this matrix.

$$\det = (40 \text{ lb})*(-1) - (20 \text{ lb})*(-1) = -20 \text{ lb}$$

Calculate *x* as the ratio of the determinant of the substituted matrix to the determinant of the coefficient matrix.

$$x = -20 \text{ lb}/-0.2 = 100 \text{ lb}$$

Substitute column vector, $\begin{bmatrix} 40 \text{ lb} \\ 20 \text{ lb} \end{bmatrix}$, into "y" positions in the coefficient matrix.

$$\begin{bmatrix} 1 & 40 \text{ lb} \\ 0.8 & 20 \text{ lb} \end{bmatrix}$$

	A	B	C	D	E	F	G	H
1				**2×2 Matrix**				
2	Equation Coefficients			Column Vector		**Det= −0.2**		=($A$3*$B$4−$A$4*$B$3)
3	1	−1		40		**x= 100**		=(D3*B4−B3*D4)/$G$2
4	0.8	−1		20		**y= 60**		=(D4*A3−A4*D3)/$G$2

Figure 1.2 Spreadsheet screenshot of Kramer's rule calculation

Calculate the determinant of this matrix.

$$\det y = 1*(20 \text{ lb}) - (0.8)*(40 \text{ lb}) = -12 \text{ lb}$$

Calculate y as the ratio of the determinant of the substituted matrix to the determinant of the coefficient matrix.

$$y = -12 \text{ lb}/-0.2 = 60 \text{ lb}$$

Figure 1.2 contains a screenshot of the solution using a spreadsheet to calculate the Kramer's rule parameters.
Solution:

$$(x, y) = (100 \text{ lb}, 60 \text{ lb})$$

All three methods produced the same result.

chapter two

Interpolation of data in tables and charts

Food engineering data will occasionally be presented in tabular form or in charts. A food scientist or engineer must be able to read tables to find values for parameters when a specific value is not given in the table. Charts are usually used for complex data whose formulas are unwieldy; the chart simplifies the extraction of data.

Tables of data are typically set up with variables along the top and left side. The needed information is then located in the body of the table. The easiest situation one encounters when reading a table occurs when the needed parameter values for both the row and column variables match exactly with those given. The data can then be read easily by noting the value at the intersection of the row and column. For example, if one wanted to know the value of the example data at parameter values of Y_2 and X_2 in Table 2.1, one would find their intersection and note that the example data value is b_2.

At times, one parameter's value will match up exactly with a column or row heading, and the other parameter's value does not have an exact match in its row or column. In this case, single interpolation must be used to calculate the data value. In other cases, interpolation must be used twice (through columns and rows) to find the value (see Tables 2.2 and 2.2a).

To find the pressure drop in a 10-m pipe with a flow rate of 2.5 m³/s, one would look down the 2.5-m³/s column in Table 2.2 until the 10-m row is reached. The cell at this intersection contains the necessary information, 1273.2 Pa.

2.1 Single interpolation

As stated earlier, interpolation is necessary when one of the table variables is not explicitly given. In Table 2.2, this would occur, for example, if information were needed on the pressure drop for a 15-m pipe. There is no entry in the table at this pipe length. However, if one knows that there is a linear relationship between pipe length and pressure drop, a value for the pressure drop could be determined. For example, we will calculate the pressure drop in a 15-m long pipe with a flow rate of 1.0 m³/s. From the Poiseuille–Hagen equation, we know that the pressure drop varies linearly with pipe length.

Table 2.1 Example Data

	Y_1	Y_2	Y_3
X_1	a_1	b_1	c_1
X_2	a_2	b_2	c_2
X_3	a_3	b_3	c_3

To interpolate from the table, we must determine three things before calculating the value of the table property.

1. Do the table extremes include the value we are interpolating? That is, does the table encompass the value that we are interested in?
2. Which two table entries surround our variable?
3. What is the fractional difference of our variable from the closest table entry?

To answer these for our example:

1. The table does include the value we are looking for (the table would include any pipe length values from 10 m to 30 m).
2. Our value of 15 m fits between the 10-m and 20-m entries.
3. Finally, to calculate the fractional difference, a simple formula is used:

Fractional difference (FD) = (value not in table – small table entry)/

(large table entry – small table entry) (2.1)

In this case, FD = (15 – 10)/(20 – 10) = 5/10 = 0.5.

To calculate the unknown table value, one virtually "backs out" this calculation for the table property.

Table property = FD (large table property – small table property)

+ small table property   (2.2)

Table 2.2 Pressure Drop (ΔP) Required to Pump Juice Through a 0.1-m-Radius Pipe with Varying Flow Rates and Pipe Length (Viscosity = 0.002 Pas)

	Flow Rate (m³/s)			
Pipe Length (m)	1.0	2.5	5.0	10.0
10	509.3	1,273.2	2,546.5	5,093.0
20	1,018.6	2,546.5	5,093.0	10,185.9
30	1,527.9	3,819.7	7,639.4	15,278.9

Here, the pressure drop for a 10-m pipe is 509.3 Pa, and for a 20-m pipe the pressure drop is 1018.6 Pa. Now, the pressure drop can be calculated for a 15-m pipe.

Pressure drop = FD'(pressure at 20 m − pressure at 10 m) + pressure at 10 m

$$= 0.5(1018.6 \text{ Pa} − 509.3 \text{ Pa}) + 509.3 \text{ Pa}$$

$$= 0.5(509.3 \text{ Pa}) + 509.3 \text{ Pa}$$

$$= 254.7 \text{ Pa} + 509.3 \text{ Pa}$$

$$= 764.0 \text{ Pa}$$

for a 15-m pipe.

This value is exactly halfway between the pressure drop values for the 10-m and 20-m pipes.

Example 2.1.1

Use Table 2.2 to calculate the pressure drop across a 30-m pipe at a flow rate of 2.0 m³/s. Answer the three fundamental questions before arriving at the answer.

Solution:

1. The table includes flow rates between 1.0 and 10.0 m³/s, so it is suitable to be used for this problem.
2. The table entries at 1.0 m³/s and 2.5 m³/s surround the flow rate of 2.0 m³/s given in the problem.
3. FD = (2.0 m³/s − 1.0 m³/s)/(2.5 m³/s − 1.0 m³/s) = (1.0 m³/s)/(1.5 m³/s) = 0.667

The pressure drop at 1.0 m³/s and 30 m is 1527.9 Pa. The pressure drop at 2.5 m³/s and 30 m is 3819.7 Pa.

Pressure drop = FD(pressure at 2.5 m³/s − pressure at 1.0 m³/s)

$$+ \text{ pressure at } 1.0 \text{ m}^3/\text{s}$$

$$= 0.667(3819.7 \text{ Pa} − 1527.9 \text{ Pa}) + 1527.9 \text{ Pa}$$

$$= 0.667(2291.8 \text{ Pa}) + 1527.9 \text{ Pa}$$

$$= 1527.9 \text{ Pa} + 1527.9 \text{ Pa}$$

$$= 3055.8 \text{ Pa}$$

This is the pressure drop for a flow rate of 2.0 m³/s in a 30-m pipe. Its value falls between the table values found for a 30-m pipe at flow rates of 1.0 m³/s and 2.5 m³/s and is closer to the value found at 30 m and 2.5 m³/s.

Table 2.3 Selected Entries from Saturated Steam Table

| Temperature (°F) | Enthalpy (Btu/lb) | |
	Saturated Liquid, h_f	Saturated Steam, h_g
60	28.06	1,087.7
70	38.05	1,092.1
80	48.04	1,096.4

Source: C.E. Power Systems.

Example 2.1.2

Use Table 2.3 to interpolate the enthalpy of saturated liquid at 66°F.

Solution:

1. This portion of the saturated steam table includes data for a range of temperatures that surround 66°F; therefore, it is acceptable to use for this problem.
2. The enthalpy values for saturated liquid (h_f) at 60°F and 70°F:
   h_f at 60°F: 28.06 Btu/lb
   h_f at 70°F: 38.05 Btu/lb
3. FD = (temperature of unknown – low temperature)/(high temperature – low temperature)

$$FD = (66°F – 60°F)/(70°F – 60°F) = 6/10 = 0.6$$

$$h = FD[h_f \text{ (at } 70°F) – h_f \text{ (at } 60°F)] + h_f \text{ (at } 60°F)$$

$$= 0.60(38.05 \text{ Btu/lb} – 28.06 \text{ Btu/lb}) + 28.06 \text{ Btu/lb}$$

$$= 0.60(9.99 \text{ Btu/lb}) + 28.06 \text{ Btu/lb}$$

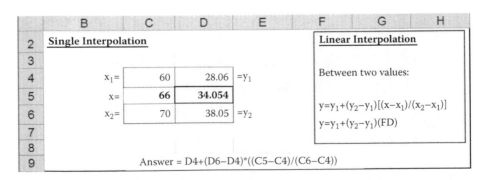

Figure 2.1 Spreadsheet screenshot of single interpolation calculation

$$= 6.00 \text{ Btu/lb} + 28.06 \text{ Btu/lb}$$

$$= 34.06 \text{ Btu/lb}$$

This enthalpy is between the enthalpies at 60°F and 70°F. A spreadsheet solution is shown in Figure 2.1.

2.2 Double interpolation

Double interpolation is necessary when both the row variable and column variable to be used are not explicitly given. The method is very similar to single interpolation. However, one must interpolate twice in one direction (either in the row variable or column variable) and once in the remaining direction. The initial interpolations are needed to give a basis for the third interpolation. The actual calculations are performed in the same manner as single interpolation.

As a simple example, calculate the pressure drop for a flow rate of 7.5 m³/s and a pipe length of 25 m using Table 2.2. Neither one of these values has a column or row explicitly dedicated to it, so double interpolation is needed. Again, the fundamental questions must be answered.

1. The table is acceptable for use because the flow rate and pipe length both fit well within the table extremes.
2. The flow rate values surrounding 7.5 m³/s are 5 m³/s and 10 m³/s, and the pipe length values surrounding 25 m are 20 m and 30 m. In this problem, interpolation will be performed twice on the pipe length variables (they are slightly easier to deal with).
3. $FD_{\text{pipe length}} = (25 \text{ m} - 20 \text{ m})/(30 \text{ m} - 20 \text{ m}) = 5 \text{ m}/10 \text{ m} = 0.50$. (This will be used for both interpolations regarding the pipe length.) By doing two interpolations over pipe length, a new table row is created, in essence, at the pipe length of 25 m (see Table 2.2a).

The first interpolation needs to be done at the lower limit flow rate (5 m³/s). The pressure drops at these conditions are 5093.0 Pa (20 m pipe length) and 7639.4 Pa (30 m pipe length).

Table 2.2a Pressure Drop (ΔP) Required to Pump Juice Through a 0.1-m-Radius Pipe With Varying Flow Rates and Pipe Length (Viscosity = 0.002 Pas)

Pipe Length (m)	Flow Rate (m³/s)				
	1.0	2.5	5.0	7.5	10.0
10	509.3	1,273.2	2,546.5		5,093.0
20	1,018.6	2,546.5	5,093.0		10,185.9
25			interpolation 1	interpolation 3	interpolation 2
30	1,527.9	3,819.7	7,639.4		15,278.9

Pressure drop (at 25 m, 5 m³/s) = $FD_{pipe\ length}$[pressure drop (at 30 m, 5 m³/s)

− pressure drop (at 20 m, 5 m³/s)]

+ pressure drop (at 20 m, 5 m³/s)

= 0.5(7639.4 Pa − 5093.0 Pa) + 5093.0 Pa

= 0.5(2546.4 Pa) + 5093.0 Pa

= 1273.2 Pa + 5093.0 Pa

= 6366.2 Pa

This value could be inserted into the interpolation 1 slot in Table 2.2a. Now, the value for the interpolation 2 slot must be calculated.

The pressure drop value at a flow rate of 10 m³/s and 20 m is 10,185.9 Pa, and at 30 m it is 15,278.9 Pa.

Pressure drop (at 25 m,10 m³/s) = $FD_{pipe\ length}$[pressure drop (at 30 m, 10 m³/s)

− pressure drop (at 20 m, 10 m³/s)]

+ pressure drop (at 20 m, 10 m³/s)

= 0.5(15,278.9 Pa − 10,185.9 Pa) + 10,185.9 Pa

= 0.5(5,093.0 Pa) + 10,185.9 Pa

= 2,546.5 Pa + 10,185.9 Pa

= 12,732.4 Pa

Since the two "table" entries surrounding the unknown value at 15 m and 7.5 m³/s are known, interpolation 3 can be calculated to finish the problem.

$FD_{flow\ rate}$ = (7.5 m³/s − 5.0 m³/s)/(10.0 m³/s − 5.0 m³/s)

= 2.5 m³/s /5.0 m³/s

= 0.50

Pressure drop (at 25 m, 7.5 m³/s) = $FD_{flow\ rate}$[pressure drop (at 30 m, 7.5 m³/s)

− pressure drop (at 20 m, 7.5 m³/s)]

+ pressure drop (at 20 m, 7.5 m³/s)

= 0.5(12,732.4 Pa − 6,366.2 Pa) + 6,366.2 Pa

= 0.5(6,366.2 Pa) + 6,366.2 Pa

= 3,183.1 Pa + 6,366.2 Pa

= 9,549.3 Pa

Does this value seem reasonable? Looking at Table 2.2, the calculated pressure drop appears to fit in the table.

Example 2.2.1

Use Table 2.2 to calculate the pressure drop in a 17.5-m pipe with a flow rate of 3.6 m³/s.

Solution:

1. The first step is to determine whether the unknown value is within the table's limits. Since the table covers flow rates between 1 and 10 m³/s and pipe lengths between 10 and 30 m, the table is appropriate for use.
2. The pipe length values surrounding 17.5 m are 10 and 20 m. The flow rate values surrounding 3.6 m³/s are 2.5 and 5 m³/s. Interpolation on the pipe length seems to be the easiest direction in which to start the problem.
3. $FD_{pipe\ length} = (17.5\ m - 10\ m)/(20\ m - 10\ m) = 7.5\ m/10\ m = 0.75$

The pressure drop values at 2.5 m³/s are 1273.2 Pa (10 m) and 2546.5 (20 m).

Pressure drop (at 17.5 m, 2.5 m³/s) = $FD_{pipe\ length}$[pressure drop (at 20 m, 2.5 m³/s)

$\qquad$ − pressure drop (at 10 m, 2.5 m³/s)]

$\qquad$ + pressure drop (at 10 m, 2.5 m³/s)

$\qquad$ = 0.75(2,546.5 Pa − 1,273.2 Pa) + 1,273.2 Pa

$\qquad$ = 0.75(1,273.2 Pa) + 1,273.2 Pa

$\qquad$ = 954.9 Pa + 1,273.2 Pa

$\qquad$ = 2,228.1 Pa

The pressure drop values at 5.0 m³/s are 5,093.0 Pa (20 m) and 2,546.5 (10 m).

Pressure drop (at 17.5 m, 5.0 m³/s) = $FD_{pipe\ length}$[pressure drop (at 20 m, 5.0 m³/s)

$\qquad$ − pressure drop (at 10 m, 5.0 m³/s)]

$\qquad$ + pressure drop (at 10 m, 5.0 m³/s)

$\qquad$ = 0.75(5,093.0 Pa − 2,546.5 Pa) + 2,546.5 Pa

$\qquad$ = 0.75(2,546.5 Pa) + 2,546.5 Pa

$\qquad$ = 1,909.9 Pa + 2,546.5 Pa

$\qquad$ = 4,456.4 Pa

Now, we are ready to interpolate the value of the pressure drop between 2.5 and 5.0 m³/s.

$$FD_{\text{flow rate}} = (3.6 \text{ m}^3/\text{s} - 2.5 \text{ m}^3/\text{s})/(5.0 \text{ m}^3/\text{s} - 2.5 \text{ m}^3/\text{s})$$

$$= 1.1 \text{ m}^3/\text{s}/2.5 \text{ m}^3/\text{s}$$

$$= 0.44$$

Using the pressure drop values surrounding the point of interest, the unknown pressure drop may be calculated.

Pressure drop (at 17.5 m, 3.6 m³/s) = $FD_{\text{flow rate}}$ (pressure drop at 17.5 m, 5.0 m³/s

$$- \text{ pressure drop at } 17.5 \text{ m, } 2.5 \text{ m}^3/\text{s})$$
$$+ \text{ pressure drop at } 17.5 \text{ m, } 2.5 \text{ m}^3/\text{s}$$

$$= 0.44 \ (4456.4 \text{ Pa} - 2228.1 \text{ Pa}) + 2228.1 \text{ Pa}$$

$$= 0.44 \ (2228.1 \text{ Pa}) + 2228.1 \text{ Pa}$$

$$= 980.4 \text{ Pa} + 2228.1 \text{ Pa}$$

$$= 3{,}208.5 \text{ Pa}$$

This pressure drop fits between the pressure drop values at 10 m and 2.5 m³/s and the values at 20 m and 5.0 m³/s. A spreadsheet solution is shown in Figure 2.2, again demonstrating how quickly software tools can generate a solution.

Example 2.2.2

Use Table 2.4 to calculate the enthalpy of superheated steam at 7 psi and 360°F.

	B	C	D	E	F	G	H
12	**Double Interpolation**						
13				**A**		**B**	
14				2.5	**3.6**	5	
15		**1st Known Value:**	10	1273.2		2546.5	
16		**Unknown Value:**	**17.5**	2228.2	3208.6	4456.4	
17		**2nd Known Value:**	20	2546.5		5093	
18							
19	**Interpolation**		Col A =E15+(E17−E15)*(($D$16−$D$15)/($D$17−$D$15))				
20	**Values:**		Col B =G15+(G17−G15)*(($D$16−$D$15)/($D$17−$D$15))				
21							
22			**Answer** =E16+(G16−E16)*(($F$14−$E$14)/($G$14−$E$14))				

Figure 2.2 Spreadsheet screenshot of double interpolation calculation

Table 2.4 Selected Entries from Superheated Steam Table (Enthalpy, Btu/lb)

Pressure (psi)	Temperature (°F)		
	300	350	400
1	1,195.7	1,218.7	1,241.8
5	1,194.8	1,218.0	1,241.3
10	1,193.7	1,217.1	1,240.6

Source: C.E. Power Systems.

Solution:

1. The abbreviated Table 2.4 contains data at conditions surrounding the needed information, so the table may be used to solve the problem.
2. Although we can work in either direction to start the problem, we will first calculate the enthalpy at the two temperatures that surround the temperature in question, 360°F. Those two temperatures are 350°F and 400°F. The pressures surrounding 7 psi are 5 psi and 10 psi.
3. $FD_{psi} = (P_{unknown\ enthalpy} - P_{lower\ table\ property})/(P_{higher\ table\ property} - P_{lower\ table\ property})$

$$FD_{psi} = (7\ psi - 5\ psi)/(10\ psi - 5\ psi) = 2\ psi/5\ psi = 0.4$$

The enthalpies at 350°F are 1218.0 Btu/lb (at 5 psi) and 1217.1 Btu/lb (at 10 psi).

h (at 7 psi, 350°F) = FD_{psi} [h (at 10 psi, 350°F) − h (at 5 psi, 350°F)]

$+ h$ (at 5 psi, 350°F)

$= 0.4(1217.1\ Btu/lb - 1218.0\ Btu/lb) + 1218.0\ Btu/lb$

$= 0.4(-0.9\ Btu/lb) + 1218.0\ Btu/lb$

$= -0.4\ Btu/lb + 1218.0\ Btu/lb$

$= 1217.6\ Btu/lb$

The enthalpies at 400°F are 1241.3 Btu/lb (at 5 psi) and 1240.6 Btu/lb (at 10 psi).

h (at 7 psi, 400°F) = FD_{psi}[h (at 10 psi, 400°F) − h (at 5 psi, 400°F)]

$+ h$ (at 5 psi, 400°F)

$= 0.4(1240.6\ Btu/lb - 1241.3\ Btu/lb) + 1241.3\ Btu/lb$

$= 0.4(-0.7\ Btu/lb) + 1241.3\ Btu/lb$

$= -0.3\ Btu/lb + 1241.3\ Btu/lb = 1241.0\ Btu/lb$

Now, we are ready for the final interpolation.

$FD_{temperature} = (T_{unknown\ enthalpy} - T_{lower\ table\ property})/(T_{higher\ table\ property} - T_{lower\ table\ property})$

$$= (360°F - 350°F)/(400°F - 350°F) = 10°F/50°F = 0.20$$

h (at 7 psi, 360°F) = $FD_{temperature}$[h (at 7 psi, 400°F) − h (at 7 psi, 350°F)]

$$+ h\ (at\ 7\ psi,\ 350°F)$$

$$= 0.2(1241.0\ Btu/lb - 1217.6\ Btu/lb) + 1217.6\ Btu/lb$$

$$= 0.2(23.4\ Btu/lb) + 1217.6\ Btu/lb$$

$$= 4.7\ Btu/lb + 1217.6\ Btu/lb$$

$$= 1222.3\ Btu/lb$$

This enthalpy appears to fit in the table at the proper position.

2.3   Interpolation in charts

Charts are used in several food engineering situations to provide data for problem solving. In most cases, the charts have become necessary because the underlying equations are complex and difficult to use. However, just as in tables, the exact data one is seeking is not always explicitly displayed in a chart. Thus, the user must be able to interpolate between data points.

The three major charts used by food engineers are (a) the psychrometric chart used to calculate the properties of air, (b) the friction factor chart (sometimes called the Moody chart or Fanning friction factor chart) used for finding the friction factor for turbulent flow in a pipe, and (c) Heisler charts used for solving transient heat-transfer problems.

Figure 2.3 contains a simple schematic diagram of the psychrometric chart. In it, one will notice there are several properties that can be determined by being able to locate a point on the chart. In this case, any two of the properties on the chart (and most psychrometric charts contain at least seven properties) can be used to specify the remaining five. Undoubtedly, the properties of interest will not lie right on a graduated line. The engineer or scientist using the chart must feel comfortable interpolating necessary values from the chart.

Interpolating in charts follows exactly from the methods of interpolation in tables. Figure 2.4 shows a more detailed schematic of a psychrometric chart, isolating three properties. In this case, suppose that the state point is specified by two parameters such as dry-bulb temperature and relative humidity, and the dew-point temperature and humidity ratio are needed. Where the horizontal line meets the saturation line, suppose there is no specific value displayed for the dew-point temperature. Interpolation must be made from the nearest graduated values for the parameter in question,

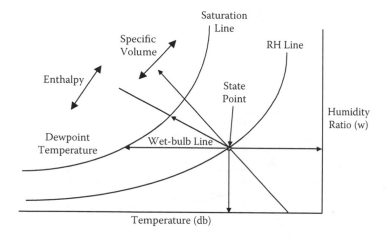

Figure 2.3 Schematic diagram of a psychrometric chart

dew-point temperature in this case. In many instances, the interpolation may be made directly on the chart. For instance, if the needed value falls halfway between two known values, then the unknown is obviously the midpoint between them. In most cases, this technique can be used for interpolating from charts. In essence, the fractional difference is being estimated visually and applied on the spot.

Because the use of charts inherently adds a degree of error associated with visually observing values, care must be used to pinpoint parameters as precisely as possible. Even things such as the thickness of a pencil lead can sometimes lead to quite different results in reading a chart. In most cases,

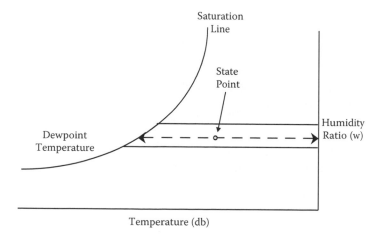

Figure 2.4 Interpolation in a psychrometric chart

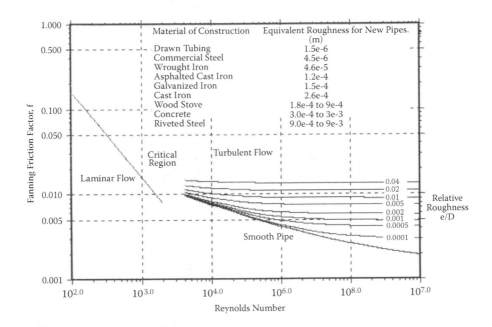

Figure 2.5 Friction-factor chart

interpolation may be made directly on the chart, especially on well-gradu-ated charts. Occasionally, access to a good chart may be limited, and inter-polation such as that discussed in sections 2.1 and 2.2 must be used. If the fractional difference were easily estimable, the solution would proceed from the estimate, as has been discussed above. In the case where the fractional difference is not easily estimable, a ruler may be used to calculate the frac-tional difference (distance between unknown and lower known value divided by the distance between the two known values). With the fractional difference calculated, the interpolation process may be followed using it and the known parameters surrounding the unknown.

An example friction-factor chart can be found in Figure 2.5. Notice the log scale on both axes. For this chart, the Reynolds number and surface rough-ness ratio are specified. The friction factor is found by drawing a horizontal line from their intersection to the left-hand axis. Care must be taken when reading the axis, but in most cases, interpolation may be made on the spot. The units for the Reynolds number, surface roughness ratio, and friction fac-tor are all dimensionless, so the chart is nonspecific for unit systems.

chapter three

Graphs and curve fitting

3.1   Coordinate systems

3.1.1   Cartesian coordinates

Graphs provide an opportunity to display data or information in a more readable form, and thus graphs are very important in the food processing industry. For example, graphs of temperature vs. time are used to determine if a food product has been properly sterilized. Also, rheological data is plotted to determine the viscous properties of foods. Plotting water content vs. time provides a visual method for analyzing drying experiments.

Graphs in Cartesian coordinates are, by far, the most common type of graph encountered. Cartesian coordinate systems in two dimensions have two axes, which are at 90° angles to one another. The intersection of the two axes produces a grid with four quadrants (see Figure 3.1). The axes may be labeled generically with x and y or, more specifically, with variables such as concentration, weight, or time. Most graphing is done in quadrant I, and one should become very familiar with creating and interpreting graphs in this quadrant.

Locating a particular point in two dimensions is done by finding the intersection created by the values on the x- and y-axes. Points are described using the x-axis variable first and the y-axis variable second. Therefore, when one wants to locate the point (4, 3), one first locates "4" on the x-axis and then "3" on the y-axis. The intersection created by these two values is the point of interest. Figure 3.2 illustrates this principle. Note that this is similar to locating values in tables.

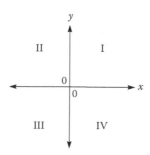

Figure 3.1 Cartesian coordinates with four quadrants

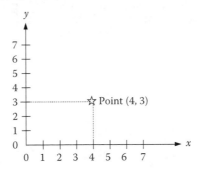

Figure 3.2 Plot of (4, 3) in Cartesian coordinates

Example 3.1.1

Plot the following set of points in Cartesian coordinates:

x	y
0	2
1	2.5
2	3
3	3.5
4	4

Solution:

Figure 3.3 is the solution.

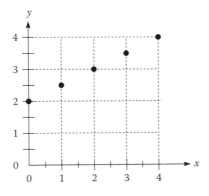

Figure 3.3 Plot of data set in Cartesian coordinates

In some cases, it is necessary to use Cartesian coordinates in three dimensions. The third dimension (normally called the z-axis) is at a 90° angle to the plane defined by the x- and y-axes. In the previous two figures, the z-axis would be coming out of the page.

3.1.2 Polar coordinates and cylindrical coordinates

Polar coordinates are another system for describing points in space. These coordinates are very useful in handling problems with circular or cylindrical objects. For example, heat or mass transfer from cylindrical or spherical food products is described using polar coordinates. The idea behind this coordinate system is that any point in two dimensions may be described by a radius and angle from the origin. The polar coordinate system is shown in Figure 3.4. The radius extends from the origin at a specified angle. The angle is described as the angle swept in a counterclockwise manner from the traditional x-axis. This angle is usually represented by units of radians (rads). Angle measurements can be converted to the familiar degree units through the following conversion:

$$180° = \pi \text{ rads} \tag{3.1}$$

so,

$$\text{rad} = \frac{180°}{\pi} \text{ degrees} \tag{3.2}$$

In general, radian units are used to describe points in polar coordinates.

The point located previously in Figure 3.2 can be located in two-dimensional polar coordinates as well. The coordinate system does not change the point; it only changes the way a point is described. Some simple relations are used to convert from Cartesian coordinates to polar coordinates. The radius is the hypotenuse of the right triangle created by the magnitudes of the x and y

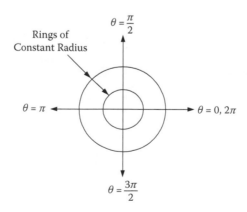

Figure 3.4 Polar coordinate system

coordinates, and the angle [radians] is measured from the *x*-axis. From geometry,

$$r^2 = x^2 + y^2 \tag{3.3}$$

$$r = \sqrt{x^2 + y^2} \tag{3.4}$$

$$\tan\theta = \frac{y}{x} \tag{3.5}$$

$$\theta = \tan^{-1}\left(\frac{y}{x}\right). \tag{3.6}$$

From equations (3.4) and (3.6), the respective radius and angle for the above point in Cartesian coordinates can be calculated for polar coordinates.

$$r = \sqrt{x^2 + y^2} = \sqrt{4^2 + 3^2} = \sqrt{16 + 9} = \sqrt{25} = 5$$

$$\theta = \tan^{-1}\left(\frac{y}{x}\right) = \tan^{-1}\left(\frac{3}{4}\right) = \tan^{-1}(0.75) = 0.64 \text{ rad}$$

Converting radian units to degrees,

$$0.64 \text{ rad} = \left(\frac{180°}{\pi \text{ rad}}\right) = 36.9°.$$

Figure 3.5 shows this graphically.

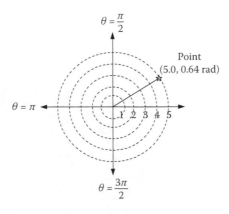

Figure 3.5 Plot of (4, 3) in polar coordinates

Example 3.1.2

Convert the point (–2, 2) in Cartesian coordinates to polar coordinates. Graph the point. Figure 3.6 shows the point plotted on polar coordinates.

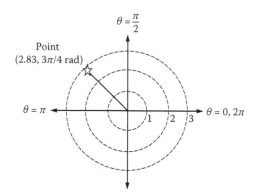

Figure 3.6 Plot of (–2, 2) in polar coordinates

Solution:

Use equations (3.2) and (3.4).

$$r = \sqrt{x^2 + y^2} = \sqrt{(-2)^2 + 2^2} = \sqrt{4+4} = 2\sqrt{2} = 2.83$$

$$\theta = \tan^{-1}\left(\frac{y}{x}\right) = \tan^{-1}\left(\frac{2}{-2}\right) = \tan^{-1}(-1) = \frac{3\pi}{4} \text{ rad}$$

To move from polar coordinates to cylindrical coordinates, one must add the third dimension, which is the same as in Cartesian coordinates. In cylindrical coordinates, it is a z-axis coming out of the page. There is no transformation necessary to convert a z-axis value from Cartesian to cylindrical coordinates. A diagram of the cylindrical coordinate system is shown in Figure 3.7.

A variety of other coordinate systems exist and are used in other applications. The principles for using them are the same as shown here. A coordinate system is simply a way of describing a point in space, and one can go back and forth between coordinate systems to find the system that most easily describes a point.

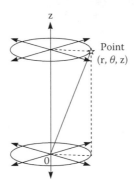

Figure 3.7 Cylindrical coordinate system

3.2 *Linear graphs*

One of the most important principles for a food scientist or engineer to master is that of plotting data and determining the equation of the best-fit line through the data. From section 3.1, one should be able to plot a set of data. This section will detail the methods one must use to find the linear equation that best fits the data. This skill is necessary so that one can calculate intermediate values of a parameter without having to interpolate. In a sense, the equation that represents data is a form of interpolation, but it is an advanced form that gives the engineer or scientist more freedom.

Recall from equation (1.15) that two constants are combined with the x and y variables to form the equation of a line. The point-slope formula is consistently the most efficient method to use to calculate the two constants. When data is linear, this method may be used under the following guidelines:

1. Two points P1 and P2 that lie on the line must be chosen.
2. The slope, m, must be calculated with equation (3.7),

$$m = \frac{y_2 - y_1}{x_2 - x_1} \tag{3.7}$$

where $P1 = (x_1, y_1)$ and $P2 = (x_2, y_2)$.
3. The y-intercept, b, is calculated by substituting one of the above points into the point-slope equation.

$$y - y_1 = m(x - x_1) \tag{3.8}$$

Rearranging,

$$y = mx + (y_1 - mx_1). \tag{3.9}$$

The constants on the right-hand side of equation (3.9) are combined and called the intercept, b.

$$b = y_1 - mx_1$$

(3.10)

Now, the equation of the line is: $y = mx + b$.

As an introductory example, let us calculate the equation of the line for the data in example 3.1.1. Looking at the plot of the data, the points appear linear. Calculation of the slope requires the selection of two points. Since the data seem to be a perfectly straight line, any two points will be satisfactory. If one is trying to find the slope of a best-fit line of data that are not perfectly linear, this luxury may not be available. For this particular example, we will choose P1 = (0, 2) and P2 = (4, 4). The slope is

$$m = \frac{y_2 - y_1}{x_2 - x_1} = \frac{4-2}{4-0} = \frac{2}{4} = \frac{1}{2}.$$

The y-intercept is

$$b = y_1 - mx_1 = 2 - \left(\frac{1}{2}\right)(0) = 2.$$

Therefore, the equation of the line is

$$y = \left(\frac{1}{2}\right)x + 2.$$

There are two different methods to use for calculating the best-fit line through data that might not be perfectly linear: the graphical method and the statistical method. The first method gives a good approximation of the equation of the line that best fits the data, whereas the second method gives a more precise equation. The graphical method consists of drawing the best-fit line through the data and then selecting the two points to use in slope and y-intercept equations. The more accurate statistical method requires the use of each data point in the following two equations:

$$m = \frac{\sum xy - \frac{\sum x \sum y}{N}}{\sum x^2 - \left(\frac{\sum x}{N}\right)^2}$$

(3.11)

$$b = \frac{\sum y - m \sum x}{N}$$

(3.12)

Additionally, we can determine how well the data fits a straight line. The correlation coefficient, r^2, obtained through equation (3.13), measures the deviation of the data from the linear equation. If r^2 is near 1.0 (>0.90), the linear equation fits the data quite well. However, a lower r^2 indicates that another, nonlinear equation should be used.

$$r^2 = \frac{\Sigma(y_i^* - \bar{y})^2}{\Sigma(y_i - \bar{y})^2}$$
(3.13)

where

    $\bar{y}$ = average y value
    y_i = individual y values
    y_i^* = individual y values calculated from the slope and intercept

Numerous computer software programs perform these calculations once the data have been entered. It is important that one understand these calculations. But in reality, one will rarely do them by hand in industry. The procedure for computer calculation usually follows the following steps:

1. x data and y data are listed in columns.
2. The data are marked, and the appropriate functions are used to generate the slope, intercept, and correlation coefficient, or perhaps a linear regression routine is called. The program automatically performs the calculations on the input data.
3. The regression data is interpreted. Usually the slope and the y-intercept are clearly marked. The r^2 is also given to show how well the data fits a straight line.

Figure 3.8 is a spreadsheet screenshot showing the slope, intercept, and r^2 of the previous example.

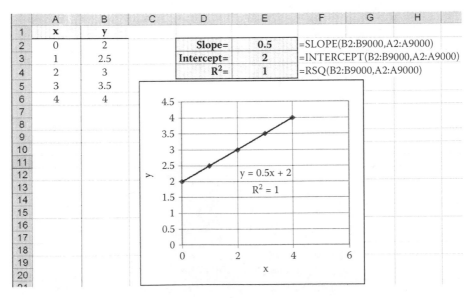

Figure 3.8 Spreadsheet screenshot showing the equation of a line

Example 3.2.1

Bingham plastic fluids may be described by the following general equation:

$$\sigma = \mu\dot{\gamma} + \sigma_0 \tag{3.14}$$

where

σ = shear stress (Pa)
μ = viscosity (Pa·s)
$\dot{\gamma}$ = shear rate (1/s)
σ_0 = yield stress (Pa)

This equation has the form of a linear equation. Calculate the best-fit line through the following data using the graphical approach and the statistical, linear regression equations (manual and computer generated). Is there a difference in the answers in this case?

σ (Pa)	$\dot{\gamma}$ (1/s)
15	0
25	2
33	4
44	6
56	8
65	10

Solution:

First, plot the data to ensure they are linear (see Figure 3.9).

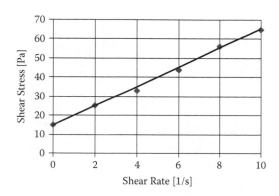

Figure 3.9 Data from example 3.2.1

One can see that the data are linear. By selecting P1 to be (10 1/s, 64 Pa) and P2 to be (0 1/s, 14 Pa), the slope and y-intercept can be calculated.

$$m = \frac{y_2 - y_1}{x_2 - x_1} = \frac{14 - 64 \text{ Pa}}{0 - 10 \text{ s}^{-1}} = \frac{-50 \text{ Pa}}{-10 \text{ s}^{-1}} = 5 \text{ Pa} \cdot \text{s}$$

The y-intercept is

$$b = y_1 - mx_1 = 14 \text{ Pa} - (5 \text{ Pa} \cdot \text{s})(0) = 14 \text{ Pa}.$$

From the graphical method

$$\sigma = (5 \text{ Pa·s}) \, \dot{\gamma} + 14 \text{ Pa}.$$

For the statistical approach, a table must be set up to handle the summations of the linear regression equations.

Shear Rate, $\dot{\gamma}$ (1/s)	Shear Stress, σ (Pa)	$\sigma, \dot{\gamma}$ (Pa/s)	$\dot{\gamma}^2$, (s^{-2})
0	15	0	0
2	25	50	4
4	33	132	16
6	44	264	36
8	56	448	64
10	65	650	100
$\Sigma\gamma = 30$ 1/s	$\Sigma\sigma = 238$ Pa	$\Sigma\gamma\sigma = 1544$ Pa/s	$\Sigma\gamma^2 = 220$ s^{-2}

Then, from equations (3.11) and (3.12),

$$m = \frac{\Sigma\dot{\gamma}\sigma - \frac{\Sigma\dot{\gamma}\Sigma\sigma}{N}}{\Sigma\dot{\gamma}^2 - \frac{(\Sigma\dot{\gamma})^2}{N}} = \frac{1544 - \frac{(30)(238)}{6}}{220 - \frac{(30)^2}{6}} = \frac{1544 - 1190}{220 - 150} = \frac{354}{70} = 5.06 \text{ Pa} \cdot \text{s}$$

$$b = \frac{\Sigma\sigma - m\Sigma\dot{\gamma}}{N} = \frac{238 - (5.06)(30)}{6} = \frac{86.29}{6} = 14.38 \text{ Pa}.$$

From the linear regression method,

$$\sigma = (5.06 \text{ Pa·s}) + 14.4 \text{ Pa}.$$

Another table must be set up to help calculate the correlation coefficient.

Shear Rate, $\dot{\gamma}$ (1/s)	Shear Stress, σ (Pa)	Calculated Shear Stress, σ (Pa)	$(\sigma_i^* - \bar{\sigma})^2$	$(\sigma_i^* - \bar{\sigma})^2$
0	15	14.4	608.4	638.4
2	25	24.52	215.1	229.4

4	33	34.64	44.4	25.3
6	44	44.76	18.8	25.9
8	56	54.88	266.8	231.4
10	65	65	641.8	641.8
$\bar{\gamma} = 5\,1/s$	$\bar{\sigma} = 39.67$ Pa		$\Sigma\,1795.3$	$\Sigma\,1792.2$

The correlation coefficient is calculated from equation (3.13) as

$$r^2 = \frac{(\sigma_i^* - \bar{\sigma})^2}{(\sigma_i - \bar{\sigma})^2} = \frac{1792.2}{1795.3} = 0.998.$$

This is a very good fit for the linear equation.

The regression output from a spreadsheet is identical to the values produced from the statistical equations. The methods produce very similar results, although one can see that the graphical method is slightly less accurate.

3.3   Logarithmic transformations

Logarithmic transformations allow for the linearization of some data that initially does not appear linear. One will encounter two types of logarithms in handling food engineering problems: the log of base 10 and the log of base e (usually referred to as the natural logarithm, or ln). These two logarithms may be used in essentially the same way, although there are times when one is preferred over the other. The basic equation of a logarithm is as follows:

$$y = 10^x \tag{3.15}$$

or

$$\log_{10} y = x. \tag{3.16}$$

Here, x is the exponent to which 10 is raised that yields the value of y. The difference between the natural log and $\log_{10}$ is the base to which x is applied. The natural log is based on e (= 2.7183) instead of 10. Using the forms of equations (3.15) and (3.16),

$$y = e^z \tag{3.17}$$

or

$$\log_e y = z. \tag{3.18}$$

From this point on, $\log_{10}$ will be referred to as "log," and $\log_e$ will be referred to as "ln."

There are three rules of logarithmic algebra that will be used frequently to rearrange equations to create more useful forms.

1. $\log(AB) = \log(A) + \log(B)$
2. $\log(A/B) = \log(A) - \log(B)$
3. $\log(A^B) = B \log(A)$

The rules apply to both log and ln. Be very careful not confuse rules 1 and 2 with something like: $\log(A) \times \log(B) = \log(A) + \log(B)$. This is a common mistake that one must be careful not to make. These rules will be used to show the simple relationship between log and ln.

Sometimes it is necessary to convert between log and ln. Equating equations (3.15) and (3.17),

$$10^x = e^z. \tag{3.19}$$

Taking the log of both sides,

$$x = z \log e = 0.4343\, z. \tag{3.20}$$

Substituting equation (3.18) into (3.20) and rearranging,

$$0.4343 \ln y = x. \tag{3.21}$$

Equating this to equation (3.16),

$$0.4343 \ln y = \log y$$

or

$$\ln y = 2.303 \log y. \tag{3.22}$$

So, one can see that the relationship between the two types of logarithms is a simple constant, 2.303.

One should become familiar and confident using logarithmic algebra. Here are a few examples to demonstrate the usefulness of these rules.

Example 3.3.1

A common problem encountered in food science and engineering that requires the use of logarithms is bacterial growth and decay. The decay of many biological organisms may be modeled with a general equation of the form

$$N = Ae^{-kt} \tag{3.23}$$

where

N = number of organisms (no./ml)
A = initial number of organisms (no./ml)

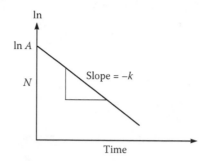

Figure 3.10 Semi-log plot of decay data to show the linear form of the equation

k = time constant affecting the rate of decay (1/s)
t = time (s).

Take the natural logarithm of both sides of this equation. Simplify the result to form an equation that has a linear form. What is the slope and intercept in this form? What variables should be plotted on the x- and y-axes?

Solution:

$$\ln N = \ln A + \ln(e^{-kt})$$

$$\ln N = -kt + \ln A \tag{3.24}$$

This equation is linear, with slope of $-k$ and y-intercept of $\ln A$. For this to take the standard linear form on Cartesian coordinates, $\ln N$ needs to be plotted on the y-axis and t needs to be plotted on the x-axis (see Figure 3.10).

Example 3.3.2

Refer to example 1.3.3 concerning the linearization of the Herschel–Bulkley equation for fluid flow.

$$\sigma = k\gamma^n + \sigma_0 \tag{3.25}$$

where

σ = shear stress in the fluid (Pa) = (N/m²)
k = fluid consistency index (s^n)
n = flow behavior index (no units)
γ = shear rate (1/s)
σ_0 = yield stress (Pa).

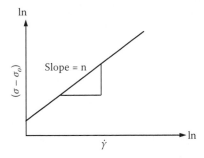

Figure 3.11 Log-log plot of rheological data to show the linear form of the equation

Use logarithmic algebra to make a linear form of this equation. How would the flow behavior index be calculated if shear stress and shear rate data were given?

Solution:

Rearrange equation (3.25) and take the ln of both sides:

$$\ln(\sigma - \sigma_o) = n \ln \dot{\gamma} + \ln k \qquad (3.26)$$

If $\ln(\sigma - \sigma_0)$ were plotted against $\ln \dot{\gamma}$ as shown in Figure 3.11, n could be calculated as the slope of this line.

3.3.1   Semi-log graphs

Figure 3.10 demonstrates that there are times when it is necessary to plot data with one logarithm variable and one normal variable in order to linearize a set of data. Graphs of this form are called semi-log plots. Using logarithmic algebra, one can calculate the form that the original equation must take. If the original equation has the form where y (or its equivalent) is equal to a constant raised to the x (or its equivalent) power, the log transformation will result in a semi-log relationship.

$$y = DC^x \qquad (3.27)$$

where C and D are constants. Taking the ln of both sides gives

$$\ln y = \ln D + x \ln C. \qquad (3.28)$$

As can be seen in Figure 3.12, ln y must be plotted against x to form a straight line with ln C as the slope.

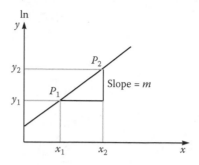

Figure 3.12 Semi-log plot demonstration

With two points P1 and P2, where $P = (x, \ln y)$,

$$m = \ln C = \frac{\ln y_2 - \ln y_1}{x_2 - x_1}. \tag{3.29}$$

Any intercept may be calculated similarly with

$$b = \ln D = \ln y_1 - mx_1. \tag{3.30}$$

One must be extremely careful when calculating slopes of nonlinear plots. There is a tendency to forget to take the log of the appropriate variables. Additionally, very few people actually plot data by hand today, and may fall into traps when plotting their data with spreadsheets or other software. Most people take their data straight to a spreadsheet and plot the data from there, and while this is very handy, one must understand the potential traps. A common mistake is to plot $\ln y$ vs. x on linear axes, and then transform the y-axis to a log scale. This is the same as taking the log twice and is incorrect. However, it is correct to plot data as either

1. $\ln y$ vs. x on linear axes **or**
2. y vs. x with the y-axis transformed to a log scale.

Regardless of which way the data is plotted, the slope and intercept are calculated with equations (3.29) and (3.30).

Example 3.3.3

The following data relate the change in number of microorganisms, N, to time of reaction, t, as discussed in example 3.3.1. Plot the data on semi-log axes, and calculate the time constant, k, which appears in equation (3.24), by using equation (3.29).

t (s)	N (no./mL)
0	100
10	36.8
20	13.5
30	5
40	1.8
50	0.7

Solution:

The data is plotted in Figure 3.13.

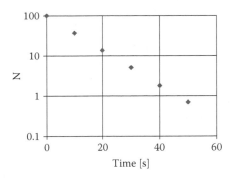

Figure 3.13 Data of example 3.3.3 plotted on semi-log coordinates

The data fits a linear form on semi-log axes. The slope, $-k$, may be calculated with equation (3.29).

$$m = -k = \frac{\ln N_2 - \ln N_1}{t_2 - t_1} = \frac{\ln(0.7) - \ln(100)}{50 - 0 \ s} = \frac{-0.357 - 4.605}{50 \ s} = \frac{-4.96}{50 \ s} = -0.099 \ s^{-1}$$

$k = 0.099 \ 1/s$

Note that the y-intercept, $N = 100$, was given as the initial number (at $t = 0$) in this problem.

3.3.2   Log-log graphs

At times, log-log transformations may be necessary to put data into a linear form. Example 3.3.2 was an example where a log-log transformation was necessary. Using the same approach as in section 3.3.1, a general equation may be developed that determines when log-log plots will produce a straight line. When the independent variable is raised to a power other than 1, a log-log transformation will produce a linear equation.

$$y = Dx^m \tag{3.31}$$

where m and D are constants. Taking ln of both sides gives

$$\ln y = \ln D + m \ln x. \tag{3.32}$$

A plot of $\ln y$ vs. $\ln x$ will produce a line with a slope of m. The slope may be calculated from equation (3.33).

$$m = \frac{\ln y_2 - \ln y_1}{\ln x_2 - \ln x_1} \tag{3.33}$$

and

$$b = \ln D = \ln y_1 - m \ln x_1 \tag{3.34}$$

where x_1 is not equal to zero. Figure 3.14 illustrates this concept.

Another more accurate method for calculating D requires taking the slope of a plot of y vs. x^m once m has been calculated from equation (3.33).

$$D = \frac{y_2 - y_1}{x_2^m - x_1^m} \tag{3.35}$$

Again, use caution when plotting data on log-log plots.

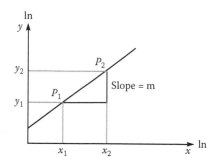

Figure 3.14 Log-log plot demonstration

Example 3.3.4

The pressure-volume relationship for an adiabatic gas is given by

$$PV^\gamma = C. \tag{3.36}$$

From the following data, determine the adiabatic expansion coefficient, γ:

V (in³)	P (psi)
54.3	61.2
61.8	49.5
72.4	37.6
88.7	28.4
118.6	18.2
194.0	10.1

Solution:

Taking the ln of both sides of equation (3.36) produces

$$\gamma(\ln P + \ln V) = \ln C. \tag{3.37}$$

Rearranging,

$$\ln P = -\gamma \ln V + \frac{\ln C}{\gamma}. \tag{3.38}$$

By taking the slope of $\ln P$ vs. $\ln V$, γ can be calculated. Before calculating, the data should be plotted to determine if it is linear (see Figure 3.15).

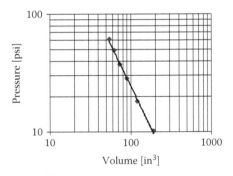

Figure 3.15 Determination of expansion coefficient

Since the data appears linear, γ can be calculated using equation (3.33).

$$-\gamma = \frac{\ln P_2 - \ln P_1}{\ln V_2 - \ln V_1} = \frac{\ln(10.1) - \ln(61.2)}{\ln(194.0) - \ln(54.3)} = \frac{-1.802}{1.273} = -1.41$$

$\gamma = 1.41$ is a typical value for gases.

chapter four

Calculus

Just the word *calculus* strikes fear in the mind of many students, but it is a subject that is very valuable to the food scientist or engineer, and one that can be mastered with a little effort. The areas of calculus that will be discussed in this chapter are differentiation and integration, along with their application to real problems in food engineering. The goal here is not to provide the material of a complete calculus text, but to provide the tools one needs to solve engineering problems. Even then, not every tool will be provided. The student should be able to take information here and use it to solve more complex problems. In some instances, one may need to consult another text to perform a difficult step, but the student should have no problem understanding the reference with the nucleus of information presented here.

4.1   Differentiation

Differentiation is the branch of calculus that allows one to calculate the rate of change of a function at any point. A function of x, $f(x)$, changes as one moves along the x-axis. Differentiation allows one to analyze the change at each specific point. Differentiation is often referred to as taking the derivative. In some cases, when $f(x)$ is simple (such as a straight line), the derivative of $f(x)$ may be a constant. Other times, it will be a new function of x (usually called $f'(x)$). The classical technique used to differentiate a function is similar to taking the slope of the function at a point. Refer to the function, $g(x)$, in Figure 4.1. If a straight-line tangent to $g(x)$ is drawn at x_1, the slope may be calculated there, and this is the value of the derivative of $g(x)$ at x_1, or $g'(x_1)$. This procedure could be repeated for each point along the function.

4.1.1   Derivatives

Refer again to Figure 4.1. Instead of taking the slope at a specific point x_1, let us take the slope at an arbitrary point, x. Let the points used to calculate the slope be $\{x, g(x)\}$ and $\{x + \Delta x, g(x + \Delta x)\}$, where Δx is small and $g(x + \Delta x)$ is considered

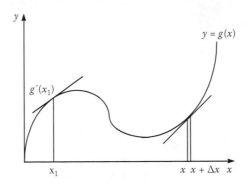

Figure 4.1 Graph of $g(x)$

to lie on the tangent line to $g(x)$ at x. The equation for the slope looks like

$$m = g'(x) = \frac{g(x + \Delta x) - g(x)}{\Delta x}. \tag{4.1}$$

Now, if Δx is considered to approach zero, the equation looks like

$$m = g'(x) = \lim_{\Delta x \to 0} \frac{g(x + \Delta x) - g(x)}{\Delta x}. \tag{4.2}$$

This becomes a very generic equation for the derivative of a function. The derivative of a function can have a variety of mathematical symbols, including: $f'(x)$ and $\frac{dy}{dx}$, where y is a function of x and $\frac{d}{dx}(y)$. We will practice using equation (4.2) with a few examples.

Example 4.1.1

Three common types of functions with which one will work are constants as well as first-order and second-order functions. The goal of this problem is to use equation (4.2) to calculate the derivative of the following functions:

1. $f(x) = 4$
2. $g(x) = 3x$
3. $h(x) = 2x^2$

Solution:

1.
$$f'(x) = \lim_{\Delta x \to 0} \frac{f(x + \Delta x) - f(x)}{\Delta x} = \lim_{\Delta x \to 0} \frac{4 - 4}{\Delta x} = 0$$

So, the slope of a constant function is zero.

2.
$$g'(x) = \frac{\lim}{\Delta x \to 0} \frac{g(x + \Delta x) - g(x)}{\Delta x} = \frac{\lim}{\Delta x \to 0} \frac{3(x + \Delta x) - 3}{\Delta x}$$

$$= \frac{\lim}{\Delta x \to 0} \frac{3x + 3\Delta x - 3x}{\Delta x}$$

$$g'(x) = \frac{\lim}{\Delta x \to 0} \frac{3\Delta x}{\Delta x} = \frac{\lim}{\Delta x \to 0} 3 = 3$$

This is the definition of the slope of a straight line, as shown previously in section 3.2.

3.
$$h'(x) = \frac{\lim}{\Delta x \to 0} \frac{h(x + \Delta x) - h(x)}{\Delta x} = \frac{\lim}{\Delta x \to 0} \frac{2(x + \Delta x)^2 - 2x^2}{\Delta x}$$

$$h'(x) = \frac{\lim}{\Delta x \to 0} \frac{2x^2 + 4x\Delta x + 2(\Delta x)^2 - 2x^2}{\Delta x}$$

$$= \frac{\lim}{\Delta x \to 0} \frac{4x\Delta x + 2(\Delta x)^2}{\Delta x} = \frac{\lim}{\Delta x \to 0} 4x + 2\Delta x$$

$$h'(x) = 4x$$

In this case, the slope (or derivative) is itself a function of x. The value of the derivative at any point is found by substituting the value of x into the derivative function.

Example 4.1.2

Calculate the rate of change of $g(x)$ and $h(x)$ above at $x = 2$ and $x = 5$. Comment on the results.

Solution:

1. $g'(x) = 3$. At $x = 2$, $g'(x) = 3$, and at $x = 5$, $g'(x) = 3$. The derivative is a constant independent of x.
2. $h'(x) = 4x$. At $x = 2$, $h'(x) = 4(2) = 8$. At $x = 5$, $h'(x) = 4(5) = 20$. The derivative depends on the value of x and is increasing with increasing x.

Very often a function will consist of first-order expressions, second-order expressions, higher-order expressions, and constants together as one function. These individual parts may be differentiated separately or together to arrive at the final answer.

Example 4.1.3

Differentiate $f(x)$ by

1. Differentiating each individual term and putting the expression back together
2. Differentiating the expression as a whole

$$f(x) = x^2 - 3x + 2$$

Solution:

1. Differentiating each individual part

$$\frac{d(x^2)}{dx} = \lim_{\Delta x \to 0} \frac{(x + \Delta x)^2 - x^2}{\Delta x} = \lim_{\Delta x \to 0} \frac{2x\Delta x + \Delta x^2}{\Delta x} = \lim_{\Delta x \to 0} 2x + \Delta x = 2x,$$

$$\frac{d(3x)}{dx} = \lim_{\Delta x \to 0} \frac{3(x + \Delta x) - 3x}{\Delta x} = \lim_{\Delta x \to 0} \frac{3x + 3\Delta x - 3x}{\Delta x}$$

$$= \lim_{\Delta x \to 0} \frac{3\Delta x}{\Delta x} = \lim_{\Delta x \to 0} 3 = 3$$

and the derivative of a constant is 0; therefore,

$$f'(x) = \frac{d(x^2)}{dx} - \frac{d(3x)}{dx} + \frac{d(2)}{dx} = 2x - 3$$

2. Differentiating the whole function

$$f'(x) = \lim_{\Delta x \to 0} \frac{f(x + \Delta x) - f(x)}{\Delta x} = \lim_{\Delta x \to 0} \frac{(x + \Delta x)^2 - 3(x + \Delta x) + 2 - x^2 + 3x - 2}{\Delta x}$$

$$f'(x) = \lim_{\Delta x \to 0} \frac{x^2 + 2x\Delta x + \Delta x^2 - 3x - 3\Delta x + 2 - x^2 + 3x - 2}{\Delta x}$$

$$= \lim_{\Delta x \to 0} \frac{2x\Delta x + \Delta x^2 - 3\Delta x}{\Delta x}$$

$$f'(x) = \lim_{\Delta x \to 0} 2x + \Delta x - 3 = 2x - 3$$

Both methods produce the same result.

4.1.2 Formal differentiation

Although equation (4.2) applies to all functions, differentiating higher-order equations is very messy and time consuming with this method. A simplifying rule (typically called the power law), which allows complex functions to be differentiated very easily, is written as

$$\frac{d(x^n)}{dx} = nx^{n-1}. \tag{4.3}$$

Each individual component of a function may be differentiated with equation (4.3) to obtain the derivative of the function. One will notice that equation (4.3) also greatly simplifies the derivatives of higher-order functions.

Example 4.1.4

Use the power law to repeat example 4.1.1.

Solution:

1. $f(x) = 4$. The exponent of x in this equation is 0. From the power law,

$$f'(x) = (4)(0)x^{(0-1)} = 0.$$

2. $g(x) = 3x$. The exponent of x is 1. From the power law,

$$g'(x) = (3)(1)x^{(1-1)} = 3x^0 = 3(1) = 3.$$

3. $h(x) = 2x^2$. The exponent of x is 2. From the power law,

$$h'(x) = (2)(2)x^{(2-1)} = 4x^1 = 4x.$$

Example 4.1.5

Calculate the derivative of the higher-order function, $g(x)$, using the power law. What is the rate of change of $g(x)$ at $x = 3$?

$$g(x) = 4x^5 - 3x^3 + x^2 - 1$$

Solution:

Each of the four terms will be differentiated sequentially.

$$d(4x^5) = (4)(5)x^4 = 20x^4$$

$$d(3x^3) = (3)(3)x^2 = 9x^2$$

$$d(x^2) = (2)x^1 = 2x$$

$$d(1) = (1)(0)x^{-1} = 0$$

The derivative of $g(x)$ is the sum of the derivatives of each term.

$$g'(x) = 20x^4 - 9x^2 + 2x$$

Evaluating the derivative at $x = 3$,

$$g'(3) = 20(3)^4 - 9(3)^2 + 2(3) = (20)(81) - (9)(9) + 6 = 1620 - 81 + 6 = 1545.$$

4.1.3   Rules of differentiation

Thus far, the basics of differentiation have been presented. Now, more information will be presented that will solidify these ideas and cover a wide range of additional topics.

Rule 1.
The derivative of a constant is zero. This has already been shown in
    example 4.1.1 (part 1).
Rule 2.
If $u(x)$ is a function of x,

$$\frac{d}{dx}(u(x))^n = n(u(x))^{n-1}\frac{d(u(x))}{dx}. \tag{4.4}$$

This is an extension of the power rule for differentiation.
Rule 3.
If u and v are functions of x,

$$\frac{d(u+v)}{dx} = \frac{du}{dx} + \frac{dv}{dx}. \tag{4.5}$$

This has been proven earlier in example 4.1.3.
Rule 4. Product Rule
Again assuming u and v are functions of x,

$$\frac{d(uv)}{dx} = v\frac{du}{dx} + u\frac{dv}{dx}. \tag{4.6}$$

The quotient rule is a special case of the product rule.

$$\frac{d(u/v)}{dx} = \frac{vdu - udv}{v^2} \tag{4.7}$$

Rule 5. Derivative of sin $u(x)$ and cos $u(x)$

$$\frac{d\sin(u)}{dx} = \cos(u)\frac{du}{dx} \tag{4.8}$$

$$\frac{d\cos(u)}{dx} = -\sin(u)\frac{du}{dx} \tag{4.9}$$

Rule 6. Derivative of ln $u(x)$

$$\frac{d(\ln u)}{dx} = \frac{1}{u}\frac{du}{dx} \tag{4.10}$$

Rule 7. Derivative of $e^{u(x)}$

$$\frac{d(e^{u(x)})}{dx} = e^{u(x)}\frac{du}{dx} \tag{4.11}$$

These rules provide the backbone for differential calculus. They do not include every detail, nor do they include every exception or special case. They do allow one a certain level of competency and provide a nucleus for solving problems.

Example 4.1.6

Use rule 2 to calculate the derivative of $y = (4x^3 - 3x^2)^3$.

Solution:

$$\frac{dy}{dx} = n(u(x))^{n-1}\frac{du}{dx} = 3(4x^3 - 3x^2)^2(12x^2 - 6x)$$

$$\frac{dy}{dx} = (36x^2 - 18x)(4x^3 - 3x^2)^2$$

Example 4.1.7

Use rules 4 and 6 to calculate the derivative of $y = x \ln x$.

Solution:

Let $u = x$ and $v = \ln x$,

$$\frac{dy}{dx} = u\frac{dv}{dx} + v\frac{du}{dx} = x\frac{1}{x}\frac{dx}{dx} + \ln x\frac{dx}{dx} = 1 + \ln x.$$

Example 4.1.8

Use rule 7 to calculate the derivative of $y = e^{-4x^2 + 2x}$

Solution:

Let $u = -4x^2 + 2x$.

$$\frac{d(e^{u(x)})}{dx} = e^{u(x)}\frac{du}{dx} = e^{-4x^2+2x}(-8x+2)$$

4.1.4   Second derivatives

Occasionally, the derivative of a previously differentiated function will be needed. This is called a second derivative, and all previous rules apply. The function (the first derivative) is differentiated just like any other function. The notation for this procedure for a function $u(x)$ is

$$f''(x) = \frac{d^2u}{dx^2} = \frac{d}{dx}\left(\frac{du}{dx}\right). \tag{4.12}$$

4.1.5   Partial derivatives

Partial derivatives occur when one wants to differentiate a function of two or more variables with respect to only one of the variables. This might occur in transient heat transfer, where the temperature is a function of position and time. Partial derivatives are straightforward, in that the portions of the function that contain the second variable are treated as constants, and the rest of the function is differentiated like any other function. The notation for partial derivatives uses ∂ in place of d.

Example 4.1.9

Calculate the partial derivative of $f(x, y)$ with respect to x:

$$f(x,y) = 4x^2y - 2xy^3$$

Solution:

$$\frac{\partial f}{\partial x} = 4y(2x) - 2y^3(1)$$

$$\frac{\partial f}{\partial x} = 8xy - 2y^3$$

The partial derivative of this function with respect to y would be different.

Check on your own to see that the partial derivative of $f(x, y)$ with respect to y is

$$\frac{\partial f}{\partial y} = 4x^2 - 6xy^2.$$

4.1.6 Applications

4.1.6.1 Time derivatives

Perhaps the most well-known and frequently used derivative is that with respect to time. The derivative of position with respect to time is velocity.

$$v(t) = \frac{ds}{dt} \tag{4.13}$$

The derivative of velocity with respect to time is acceleration.

$$a(t) = \frac{dv}{dt} = \frac{d^2s}{dt^2} \tag{4.14}$$

These can be visualized as one considers position as a function of time. The velocity is the rate of change of position; therefore, it is the first derivative of the position function. Acceleration, in turn, is the rate of change of velocity. It is the first derivative of velocity or the second derivative of position.

Example 4.1.10

Given the position function, $s = 8t^3 - 4t^2 + 2t - 1$ (m). What is the velocity and acceleration at $t = 3$ s?

Solution:

$$v = \frac{ds}{dt} = 24t^2 - 8t + 2$$

$$v(3) = 24(3)^2 - 8(3) + 2 = 24(9) - 24 + 2 = 194 \text{ m/s}$$

$$a = \frac{d^2s}{dt^2} = 48t - 8$$

$$a(3) = 48(3) - 8 = 144 - 8 = 136 \text{ m}^2/\text{s}$$

4.1.6.2 Maxima and minima

Differentiation may be used to find a maximum or minimum point of a function. At times, one will want to know what value of the independent variable will produce the maximum or minimum value in the function. One could plot the function and determine where the function has a minimum or maximum. Another method utilizes differentiation.

The method for determining a maximum or minimum value consists of four steps:

1. Evaluate the function at its upper and lower limit for the independent variable. Sometimes the function will be increasing or decreasing over an entire range of values for the independent variable. This step requires that one check the limits of the independent variable range to determine the function's values there.
2. Take the derivative of the function with respect to the independent variable. Set the derivative equal to zero and solve for the independent variable, which makes the derivative equal to zero. This value is a maximum or minimum and is called a critical point.
3. Take the second derivative of the function. Solve for the value of the second derivative at the maximum or minimum. If the second derivative is positive at this point, the point is a minimum; if the second derivative is negative at this point, the point is a maximum.
4. Compare the maximum or minimum point with the points from the upper and lower limits to determine which is the true maximum or minimum.

As an example, consider an equation for thermal conductivity, which is dependent on temperature. The thermal conductivity, k, will be at a maximum at a certain temperature. A food processor may want to operate at that temperature to maximize the rate of heat transfer.

$$k_w = 0.57109 + 0.0017625T - 6.7306(10)^{-6}T^2 \qquad (4.15)$$

where T is in °C and k_w is in W/m². Find the temperature that yields the maximum k_w over the range of temperature from 50°C to 150°C.

1. Check the upper and lower limits.

$$k_w(50°C) = 0.57109 + 0.0017625(50°C) - 6.7306(10)^{-6}(50°C)^2$$

$$= 0.6424 \text{ W/m}^2$$

$$k_w(150°C) = 0.57109 + 0.0017625(150°C) - 6.7306(10)^{-6}(150°C)^2$$

$$= 0.6840 \text{ W/m}^2$$

2. Take the derivative and set it equal to zero. Solve for the temperature.

$$\frac{dk_w}{dT} = 0.0017625 - 1.3461(10)^{-5}T = 0$$

$$T = \frac{0.0017625}{1.3461(10)^{-5}} = 130.9°C$$

3. Take the second derivative and insert the value at the critical point.

$$\frac{d^2 k_w}{dT^2} = -1.3461(10)^{-5}$$

This is always negative, so the temperature at the critical point is a maximum. The value for k_w at this point is

$$k_w(130.9°C) = 0.57109 + 0.0017625(130.9°C) - 6.7306(10)^{-6}(130.9°C)^2 = 0.6865 \text{ W/m}^2.$$

This value is greater than both of those at the upper and lower limits.

4.2 Integration

Integration is the opposite of differentiation. In this case, one knows the derivative of a function and wants to find the equation for the function that fits the derivative. This operation has many applications and uses, including calculating the area underneath a curve. That will be the starting point of this discussion.

4.2.1 The antiderivative

Integration can be thought of as working backward through differentiation. That is, one is looking for the function that will produce the derivative shown upon differentiation. Equation (4.16) shows the mathematical relationship where $F(x)$ is the desired function and $f(x)$ is given.

$$\frac{dF(x)}{dx} = f(x) \tag{4.16}$$

In some simple cases, $F(x)$ can be calculated from our previous experience. For example, if $\frac{dy}{dx} = B$, where B is a constant, we know that $y = Bx$ or $y = Bx + C$ (another constant).

The original function, y, might also have some constants in it that disappear when it is differentiated. So, the proper antiderivative, or integral, is $Bx + C$, where C is another arbitrary, unknown constant. The easiest way to check to see if the antiderivative is correct is to differentiate it and determine if the differential is equivalent to the original expression. There are really no specific rules to finding an antiderivative; one must rely on intuition and experience. Formal integration, in the next section, will provide a more rigorous approach.

Example 4.2.1

Find the antiderivatives for the following functions:

1. $\dfrac{dy}{dx} = 5$

2. $\dfrac{dy}{dx} = 4x$

3. $\dfrac{dy}{dx} = 2x - 1$

Solution:

1. As shown previously, the antiderivative of a constant is a linear function of x.

$$y = 5x + C$$

   The differential of this expression is 5, so it must be an appropriate answer.

2. This is a more complex function than a constant. The derivative of x^2 is $2x$. The antiderivative of y should be a multiple of x^2.

$$y = 2x^2 + C$$

   Again, the arbitrary constant, C, must be included to account for any missing terms.

3. This combines two degrees of order. The antiderivative is calculated by combining the techniques and reasoning from parts 1 and 2 of this example.

$$y = x^2 - x + C$$

   Again, when this is differentiated, the original expression is obtained.

4.2.2 Formal integration

As the name implies, this section will detail the rules of integration in a more formal way. The lessons learned in finding antiderivatives help to build the basic rules of integration. One might have noticed that, when differentiating a function, the resulting derivative is always of a lesser order (lower exponent) than the original function by one degree. Integration, on the other hand, always produces a function of a higher order (higher exponent) by one degree than the original starting function. The rules used to develop integration follow almost identically the rules for differentiation.

4.2.3 Rules of integration

Before beginning with the rules of integration, the integral notation must be presented. Much like $\frac{du}{dx}$ is used to note differentiation of the function $u(x)$, $\int u dx$, denotes that the function $u(x)$ inside will be integrated with respect to x.

Rule 1.

$$\int x^n dx = \frac{1}{n+1}x^{n+1} + C \quad \text{for} \quad n \neq -1 \tag{4.17}$$

Rule 2.

A constant in an integral may be moved outside of the integral without affecting the outcome.

$$\int cf(x)dx = c\int f(x)dx \tag{4.18}$$

Rule 3.

If u and v are functions of x,

$$\int (du + dv) = \int du + \int dv. \tag{4.19}$$

Rule 4. Integral of sin $u(x)$ and cos $u(x)$
When u is a function of x, $u(x)$, then

$$\int \sin(u(x))du = -\cos(u) + C \tag{4.20}$$

$$\int \cos(u(x))du = \sin(u) + C. \tag{4.21}$$

Rule 5.

$$\int \frac{du}{u} = \ln u + C \tag{4.22}$$

Rule 6. Integral of $e^{u(x)}$

$$\int e^{au}du = \frac{1}{a}e^{au} + C \tag{4.23}$$

These rules can be used together and/or separately to solve a variety of problems.

Example 4.2.2

Use integration rule 1 to integrate $\frac{dy}{dx} = 3x^2 - 4x + 5$.

Solution:

Multiplying both sides by dx gives,

$$dy = (3x^2 - 4x + 5)dx.$$

Integrating both sides,

$$\int dy = y$$

$$\int (3x^2 - 4x + 5)dx = 3\int x^2 dx - 4\int x dx + 5\int dx = \frac{3}{3}x^{2+1} - \frac{4}{2}x^{1+1} + 5x + C$$

$$\int (3x^2 - 4x + 5)dx = x^3 - 2x^2 + 5x + C$$

$$y = x^3 - 2x^2 + 5x + C.$$

4.2.4 Closed integrals

Closed integrals, sometimes called definite integrals, are solved in much the same manner as open integrals. A closed integral is one that has an upper limit and a lower limit. These limits are noted on the integral sign. When these limits are present, the problem is solved in a slightly different manner than open integrals. The function is integrated as usual, but the solution is then evaluated at the upper and lower limits. The final answer is the difference between the integral evaluated at the upper limit and the integral evaluated at the lower limit.

Example 4.2.3

Evaluate the following closed integral:

$$y = \int_{2}^{3} 3x^2 dx$$

Solution:

$$y = \int_{2}^{3} 3x^2 dx = \frac{3}{2+1}x^{2+1}\Big|_{2}^{3} = x^3\Big|_{2}^{3} = (3)^3 - (2)^3 = 27 - 8 = 19$$

Example 4.2.4

Evaluate the following closed integral:

$$y = \int_{1}^{5} e^{x/10} dx$$

Solution:

From integration rule 6, the integration should be carried out as follows:

$$y = \int_1^5 e^{x/10} dx = 10 e^{x/10}\Big|_1^5 = 10(e^{5/10} - e^{1/10}) = 10(e^{0.5} - e^{0.1}) = 5.436$$

4.2.5 Trapezoidal rule

An important application of definite integrals is their use in calculating the area under a curve. The trapezoidal rule is a technique that allows one to approximate the area underneath a curve between two points. Figure 4.2 shows a curve, and we would like to calculate the area between points a and b. The trapezoidal rule allows one to approximate this area by creating small trapezoids and summing the area occupied by them.

The area of a trapezoid is calculated from the area of a rectangle with height $\left(\frac{y_n + y_{n+1}}{2}\right)$ and width $\Delta x = x_{n+1} - x_n$,

$$A = \left(\frac{y_n + y_{n+1}}{2}\right)(x_{n+1} - x_n) = \left(\frac{y_n + y_{n+1}}{2}\right)\Delta x. \tag{4.24}$$

The width of each trapezoid is found by

$$\Delta x = \frac{b - a}{n} \tag{4.25}$$

where n is the number of increments chosen between a and b.

The area under the curve can be determined by summing the individual area of each trapezoid. As one might guess, as the number of trapezoids increases (as n becomes larger), the accuracy of the method increases. The

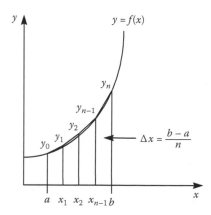

Figure 4.2 Trapezoids to approximate the area under $f(x)$

summation of trapezoidal areas can be simplified a little through the use of the trapezoidal rule. This rule states that

$$A_{total} = \left(\frac{b-a}{2n}\right)(y_0 + 2y_1 + 2y_2 + \cdots + 2y_{n-1} + y_n). \tag{4.26}$$

As n increases, A_{total} becomes more accurate.

Example 4.2.5

Use the trapezoidal rule to approximate the area under the curve $y = x^2 + 1$ from $a = 1$ to $b = 3$. Use $n = 4$.

Solution:

First, calculate Δx.

$$\Delta x = \frac{b-a}{n} = \frac{3-1}{4} = \frac{2}{4} = 0.5$$

Second, set up a table for x and y values at each increment of Δx,

x	y
1	2
1.5	3.25
2	5
2.5	7.25
3	10

Lastly, plug these values into equation (4.26).

$$A_{total} = \left(\frac{b-a}{2n}\right)(y_0 + 2y_1 + 2y_2 + 2y_3 + y_4) = \left(\frac{\Delta x}{2}\right)(y_0 + 2y_1 + 2y_2 + 2y_3 + y_4)$$

$$A_{total} = (0.25)(2 + (2)(3.25) + (2)(5) + (2)(7.25) + 10) = (0.25)(2 + 6.5 + 10 + 14.5 + 10)$$

$$A_{total} = (0.25)(43) = 10.75$$

4.2.6   Areas by integration

The area under a curve may be calculated exactly by integrating the equation of the curve over the limits of integration. The trapezoidal rule provides a good starting point. If the number of trapezoids were increased to infinity, the width of each trapezoid would decrease to an infinitesimal value, dx. The height of the trapezoid could be approximated by the value of the function at the point $y(x)$ instead of the average of two points. When these are put together, they give

$$A = y(x)dx. \tag{4.27}$$

Integration over the upper and lower limits serves as the summation step so that the total area under the curve can be calculated as

$$A_{total} = \int_a^b y(x)dx. \tag{4.28}$$

This concept is utilized in many areas of food science, including organic chemistry, quantification of chemicals, statistics, and chemical reaction kinetics.

Example 4.2.6

Repeat example 4.2.5 using integration to calculate the area of $y = x^2 + 1$ from 1 to 3. Comment on any differences.

Solution:

Write the area in an integral form.

$$A_{total} = \int_1^3 (x^2 + 1)dx = \int_1^3 x^2 dx + \int_1^3 dx = \frac{1}{3}x^3 + x \Big|_1^3 = \left[\left(\frac{1}{3}\right)(3)^3 + 3\right] - \left[\left(\frac{1}{3}\right)(1)^3 + 1\right]$$

$$A_{total} = (9 + 3) - \left(\frac{1}{3} + 1\right) = 12 - \frac{4}{3} = \frac{32}{3} = 10.667$$

One can see from this example that the exact solution is 10.6666. The trapezoidal rule with only four trapezoids approximated the area to be 10.75. The accuracy in this case is quite good, although, at other times, the results might not be so close.

4.3 Differential equations

Differential equations such as $\frac{dx}{dt}$ or $\frac{dy}{dx}$ are equations that contain a differential element as part of the equation. To solve first-order differential equations (first derivatives), a single integration step is required with the substitution of boundary conditions. More complicated differential equations (second- or higher-order differentials) require two or three integration steps and numerous boundary conditions. When we speak of order in describing a differential equation, we are speaking of the single term in the equation that has been differentiated most. First-order equations have a term that has been differentiated one time; second-order equations have a term that has been differentiated twice (second derivative) and may contain other lesser-order terms, and so forth.

Boundary conditions are the guidelines by which the problems must be solved. A boundary condition sets a limit on the value of the dependent variable at a certain independent variable. One boundary condition is required for each degree of differentiation. If one has a second-order term (derivative) in the equation, two boundary conditions are necessary to solve the equation. Initial conditions are used in problems that have differentials with respect to time. These conditions state the value of the dependent variable at time zero.

The simplest differential equations are those that contain one first-order differential element and that are separable. By separable, we mean that the two variables involved in the equation may be separated to either side of the equal sign so that only functions of one variable are on each side of the equal sign. A common heat-transfer problem in Cartesian coordinates where heat is transferred in one direction through a slab involves the solution of Fourier's law,

$$\frac{\dot{q}}{A} = -k\frac{dT}{dx} \tag{4.29}$$

where

$\dot{q}$ = rate of heat transfer (W)
A = slab cross-sectional area (m²)
k = thermal conductivity (W/m·K)
T = temperature (K)
x = coordinate in the direction of heat flow (m).

The area of heat transfer, A, is perpendicular to x and equal to $L * W$. Therefore, it is not a function of x and may be considered constant. The thermal conductivity, k, may also be considered constant in most cases where the temperature change is not too great. To solve this first-order differential equation, separate dT and dx to opposite sides of the equal sign.

$$\frac{\dot{q}}{A}dx = -kdT \tag{4.30}$$

Since $\dot{q}$, A, and k are constants, both sides may be integrated as closed integrals with the boundary conditions at $x = x_1$, $T = T_1$, and at $x = x_2$, $T = T_2$.

$$\frac{\dot{q}}{A}\int_{x_1}^{x_2} dx = -k\int_{T_1}^{T_2} dT \tag{4.31}$$

$$\frac{\dot{q}}{A}(x_2 - x_1) = -k(T_2 - T_1) = k(T_1 - T_2) \tag{4.32}$$

$$\dot{q} = kA\frac{(T_1 - T_2)}{(x_2 - x_1)} \tag{4.33}$$

Equation (4.33) is the familiar form of Fourier's law in Cartesian coordinates.

Example 4.3.1

Solve Fourier's law for heat conduction through the shell of a hollow sphere of radius, r_2. The differential equation is

$$\frac{\dot{q}}{A} = -k\frac{dT}{dr}. \tag{4.34}$$

The temperature at the inside surface ($r = r_1$) is T_1. The temperature at the outside surface ($r = r_2$) is T_2. Solve for an expression for $\dot{q}$.

Solution:

Let us separate the differential elements first.

$$\frac{\dot{q}}{A}dr = -kdT \tag{4.35}$$

Before integrating this, we must decide if $\dot{q}$, A, and k are constants. Again, $\dot{q}$ and k are constants, but A, the surface area of the sphere, is $4\pi r^2$. This must be substituted into equation (4.35) before integrating.

$$\frac{\dot{q}}{4\pi}\frac{dr}{r^2} = -kdT \tag{4.36}$$

Using similar boundary conditions as before (at $r = r_1$, $T = T_1$ and at $r = r_2$, $T = T_2$), the equation may be integrated.

$$\frac{\dot{q}}{4\pi}\int_{r_1}^{r_2}\frac{dr}{r^2} = -k\int_{T_1}^{T_2}dT \tag{4.37}$$

$$\frac{\dot{q}}{4\pi}\left(\frac{1}{-1}\right)\left(\frac{1}{r}\right)\Big|_{r_1}^{r_2} = -k(T_2 - T_1) = k(T_1 - T_2) \tag{4.38}$$

$$\frac{\dot{q}}{4\pi}\left(\frac{1}{r_1} - \frac{1}{r_2}\right) = k(T_1 - T_2) \tag{4.39}$$

A final rearrangement gives

$$\dot{q} = 4\pi k\frac{(T_1 - T_2)}{\left(\dfrac{1}{r_1} - \dfrac{1}{r_2}\right)}. \tag{4.40}$$

This is slightly different than the expression for $\dot{q}$ in a flat slab.

A final application in this section is initial-value problems. Unsteady-state heat transfer, mass transfer, and fluid flow problems fit this category, as well as many others. Consider a hot steel cube immersed in cool water. In this situation, the steel cube will conduct heat quickly. The following equation is used to describe the heat transfer

$$\dot{q} = hA(T_\infty - T) = c_p \rho V \frac{dT}{dt} \tag{4.41}$$

where

T_∞ = temperature of the fluid (K)
T = temperature (K) of the cube at any time, t (s)
V = volume of the cube (m^3)
A = surface area (m^2)
h = convective heat-transfer coefficient (W/m^2·K)
c_p = specific heat of the cube (J/kg·K)
ρ = density of the object (kg/m^3).

The initial condition for this problem is that at $t = 0$, $T = T_0$, some initial temperature.

Now, the constants are arranged together and the variables separated to form

$$\frac{hA}{c_p \rho V} dt = \frac{dT}{T_\infty - T}. \tag{4.42}$$

Integrating this with the initial condition gives

$$\frac{hA}{c_p \rho V} \int_0^t dt = \int_{T_0}^T \frac{dT}{T_\infty - T} \tag{4.43}$$

$$\frac{hA}{c_p \rho V} t = -\ln(T_\infty - T)\Big|_{T_0}^T = -[\ln(T_\infty - T) - \ln(T_\infty - T_0)] = -\ln\left(\frac{T_\infty - T}{T_\infty - T_0}\right) = -\ln\left(\frac{T - T_\infty}{T_0 - T_\infty}\right).$$

$$\tag{4.44}$$

Now, the temperature distribution at any time may be calculated with

$$\left(\frac{T - T_\infty}{T_0 - T_\infty}\right) = e^{-\left(\frac{hA}{c_p \rho V}\right)t}. \tag{4.45}$$

These problems just begin to show the subject of differential equations. Problems vary in complexity to many extremes. From this information, one should be able to solve any first-order separable equation and, hopefully, understand other reference texts when seeking to solve more difficult problems.

chapter five

Problem solving

In food engineering, and any other mathematically-oriented discipline, one must become adept at solving problems using equations and mathematical principles. Many of the problems that food scientists and engineers face require the ability to apply physical and mathematical principles to solve equations and calculate necessary information. Thus, problem-solving abilities are extremely important, and we should be familiar with the general steps involved in problem solving. In many circumstances, word problems are used to describe a problem. To solve a problem, one must

- Understand the intent of the problem
- Understand what information is given and what is needed
- Decide which approach is required
- Translate the statement into a mathematical expression
- Solve the problem

Any of these steps can inhibit one's ability to solve a problem.

Before we talk about problem-solving strategies, we should understand the several types of problems that may be encountered in food engineering. These are:

Simple, close-ended problems: These are problems with a single correct answer that can be solved using a single correct approach. Manipulation of mathematical equations falls into this category. For example, given the required information on flow rates and heat-transfer coefficients, one can calculate the temperature of a product exiting a heat exchanger using the appropriate equations for heat transfer.

Complex, close-ended problems: Sometimes problems have a single correct answer, but that answer can be found using more than one approach. These are complex, close-ended problems, typical of troubleshooting problems often found in the food industry. For example, if a piece of glasslike material is found in a food product, there is a single correct identification, although several different techniques might be utilized to identify the material.

Complex, open-ended problems: The highest level of problem is one where there are multiple correct answers with multiple approaches. A design problem falls into this category. For example, if your boss asks you to develop a low-fat version of an existing product, there may be many potential solutions to this problem. The "correct" answer in this case depends on the situation.

There are many books that describe the problem-solving process. However, they all follow basically the same steps. Recognize, however, that problem solving is not a linear process. We often follow a complex path from problem statement to solution, with many starts and stops, and numerous cycles back to previous steps. Despite the nonlinearity of problem solving, it is still valuable to think about the general steps that we all use in some form or another. A simple plan for solving problems involves the following steps:

1. *Motivation*: Understand the importance of the problem. Why is it necessary to solve this problem? What will I learn? Remember that problem-solving skills are critical for success in all aspects of life, not just in your career in the food industry. Good problem solvers are more successful in all situations.

2. *Define the problem*: Clearly understand the words that are written about the problem and what type of problem you are facing. Is this a problem on mass balances or energy balances, or both? Probably the most important step in solving a problem is translating the words into either pictures or mathematical expressions, or both. Can you draw a diagram that represents the problem? If so, label all the components, label the known and unknown parameters, and clearly define the desired parameters to be found. We cannot stress strongly enough the importance of drawing diagrams as an aid in solving problems.

3. *Think about the problem*: Here is where you should be thinking about exactly what the problem entails. What are you really being asked to do? What principles might apply? Do you have enough information or too much information? Can you attack this problem with what you already know? If not, what additional information will you need? Where will you find this information? This is an important part of problem solving, although one that is often neglected. If you jump right from the problem statement to a solution plan, then you might be missing something important, and wind up having to go back to the beginning.

4. *Plan the solution*: Most problem-solving books recommend that you spend some time planning the solution strategy before you actually implement it. Do you have all the necessary information and resources? If not, then you must go back through steps 2 and 3 until you reach the point where you are ready to solve the problem. In more complex problems, it may be necessary to break a problem down into smaller, more easily managed subproblems, and work on each of them individually for a little while.

5. *Implement the solution strategy*: Finally, after you have thought about the problem, have made sure it is well defined, and are confident that

you have the appropriate strategy, then you can solve the problem. Many of us jump right to this step first without going through the previous steps and wind up being unable to continue. At that point, it is easy to give up because the problem is not easily solved. More experienced problem solvers know that you must often go around the problem-solving process a time or two until you finally reach the point where you can solve it.

6. *Check the solution*: It is always a good idea to look at your solution and ask yourself if this sounds correct. Have you calculated a heat exchanger that must be 2 miles long? If so, then your solution is probably incorrect, and you need to check all assumptions and calculations.

7. *Generalize the solution strategy*: Have you truly learned anything new about solving problems? It is always a good idea to look back over your approach and look for ways to improve or expand upon what you have done. Can you generalize your approach to other problems?

Now let us utilize these steps to solve a problem.

Example 5.1

How much water must be added to a concentrated cleaning solution to attain 100 lb of cleaner at the proper strength. Also, how much of the initial concentrate is required? The concentration of cleaning agent of the high-strength cleaner is 50%, and the desired concentration of cleaning agent in the final product is 12%.

Solution:

1. *Motivation*: This problem is one that is typically faced in the food industry, since cleaners are normally shipped in concentrated form. When a clean-in-place system is filled with cleaner, water must be added to the concentrate to attain the proper strength. Thus, this is a problem that has definite industrial importance.

2. *Define the problem*: We are asked to calculate the amount of water to be added to a concentrate to make 100 lb of cleaning solution. Obviously this will require a mathematical solution. It may not be immediately obvious, but this is a mass-balance problem, since we have to make sure that the amount of the active cleaning ingredient is conserved. We can draw a diagram that defines this process (Figure 5.1). Two materials, water and concentrated cleaning solution, are added to produce 100 lb of cleaning solution. Let us label the amount of water needed as W and the amount of concentrated cleaner needed as C. We also know the concentration of cleaning agent in our two feed materials (0% for water and 50% for concentrate) and our final product (12% concentrate).

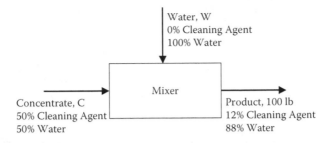

Figure 5.1 Schematic diagram of mixer

3. *Think about the problem*: As mentioned above, this is a mass-balance problem. We will need to calculate how much water and concentrated cleaner must be added to produce 100 lb of a 12% cleaning solution. How much water and cleaning agent are required in the final solution? Since mass must be conserved, this must be the same as the amount added.

4. *Plan the solution*: Mass-balance problems require us to track both total mass of material as well as the mass of individual components. In this problem, we have two components: water and cleaning agent. Thus the water entering the mixer in our diagram must equal the amount of water in our final 12% cleaning solution. In the same way, the amount of cleaning agent entering our mixer must equal the amount of cleaning agent in our final 12% cleaning solution. In the final cleaning solution, we have (0.12 * 100 lb) or 12 lb of cleaning agent and (100 − 12 lb) or 88 lb of water. Thus, we can perform both a balance on the total amount of material added and a balance on one of the components, either water or cleaning agent.

5. *Implement the solution strategy*: Writing a mass balance on the total amount of material added gives

mass of water + mass of concentrate = mass of final cleaning solution

$$W + C = 100 \text{ lb.} \tag{5.1}$$

A mass balance on the amount of cleaning agent gives

mass of cleaning agent in water + mass of cleaning agent in concentrate

= mass of cleaning agent in final cleaning solution

$$(0) \, W + (0.5) \, C = (0.12) \, 100 = 12 \text{ lb.} \tag{5.2}$$

Solving equation (5.2) gives

$$C = 12/(0.5) = 24 \text{ lb.}$$

Substituting $C = 12$ lb into equation (5.1) and solving for W gives

$$W = 100 - C = 100 - 24 = 76 \text{ lb.}$$

Thus, we need to mix 24 lb of concentrated cleaner with 76 lb of water to produce a 12% cleaning solution.

6. *Check the solution*: Are these numbers reasonable? They seem reasonable, since they are both less than 100 lb and clearly add up to 100 lb. In this case, however, we can check the answer by performing a mass balance on water.

mass of water added + mass of water in concentrate

$$= \text{mass of water in final cleaning solution}$$

$$= (1.0)W + (0.5)C$$

$$= (0.88)\ 100 \text{ lb} = 88 \text{ lb} \tag{5.3}$$

Substituting $C = 24$ lb of concentrate and 76 lb of water gives

$$76 \text{ lb} + (0.5)\ 24 \text{ lb} = 88 \text{ lb of water.}$$

Thus, we have checked our solution and found it be correct. It is a good idea to check your answers whenever possible.

7. *Generalize the solution strategy*: How can we use this solution strategy in other situations? For example, suppose we were given 100 lb of concentrate and asked to calculate how much cleaner we could make and how much water would be required? The solution strategy we followed would allow us to solve this problem as well. Where else might this problem be applied?

chapter six

Gases and vapors

Many food engineering operations make use of gases and vapors, and their behavior is sufficiently complex to merit a section here. The ideal gas law is quite straightforward. However, when we begin to use it, we often are confronted with a bewildering array of pressures (for example, atmospheric, absolute, gauge, partial, vapor). In what follows, we try to make the distinctions among the various forms of pressure.

When solving problems, it is important to understand the behavior of water, which can exist in the gaseous state at ambient temperature, even though under this condition the normal state of water is liquid. This is true because the boiling point of water (100°C) is not too much greater than ambient temperature. Permanent gases (for example, O_2, N_2, CO_2), on the other hand, must be compressed significantly to liquefy them, because they have boiling points much lower than ambient temperature. We can define gases and vapors more specifically. A permanent gas refers to the gaseous state of a molecule that has a boiling point much below ambient temperature. A vapor refers to the gaseous state of a molecule that has a boiling point not too much different than ambient temperature.

In what follows, we will try to make clear these and other issues that are important when solving problems involving gases and vapors.

6.1 Pressure

Figure 6.1 shows the relationship between absolute (P_{abs}), atmospheric (P_{atm}), and gauge (P_{gauge}) pressures. Strictly speaking, atmospheric pressure itself is an absolute pressure. It can be said that gauge pressure is equal to the difference between two absolute pressures. Given the U-tube fluid manometers depicted in Figure 6.1, P_{gauge} can be calculated from the height difference (Δh) of the manometer fluid in the two legs of the U-tube. Also, given that pressure is equal to a force per unit area, it can be shown that the magnitude of gauge pressure is

$$P_{gauge} = \rho g \Delta h \qquad (6.1)$$

where ρ is manometer fluid density and g is acceleration of gravity (980 cm/s^2; 32 ft/s^2).

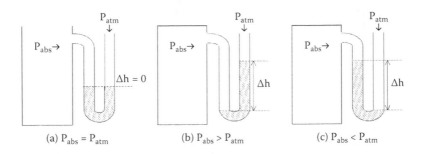

Figure 6.1 Relationships between absolute (P_{abs}) and atmospheric (P_{atm}) pressure demonstrated with a U-tube manometer

In Figure 6.1(a), there is no height difference, so $P_{gauge} = 0$, and the absolute pressure inside the chamber is the same as the atmospheric pressure outside. Usually gauges, even mechanical ones, are set so that gauge pressure is zero when absolute pressure is equal to atmospheric pressure. In Figure 6.1(b) $P_{abs} > P_{atm}$ because the level of the manometer fluid on the absolute side of the U-tube is lower than that on the atmospheric side (the manometer fluid is pushed toward the atmospheric side of the manometer). We can now say:

$$P_{abs} = P_{atm} + P_{gauge} \tag{6.2}$$

Note that this is equivalent to saying $P_{gauge} = P_{abs} - P_{atm}$. Equation (6.2) is relevant to the situation where steam would be pumped into a retort. With no steam in the retort, we would have the situation of Figure 6.1(a). Once steam is added to the retort, equation (6.2) can be used to calculate absolute pressure in the retort.

In Figure 6.1(c), the absolute pressure is less than the atmospheric pressure. This means there is a vacuum. It is common to say that a vacuum has been pulled, because vacuum pumps are used to remove or pull gas molecules out of chambers. The magnitude of the vacuum is equal to the magnitude of P_{gauge}. It is common to express the magnitude of a vacuum positively, often as inches (in.) or millimeters (mm) of mercury (Hg). If external pressure is 1 atm, the maximum possible vacuum has the magnitude of 1 atm (29.92 in. Hg; 760 mm Hg). Since P_{gauge} is equal to the difference between two absolute pressures, we can say,

$$P_{vac} = -P_{gauge} = P_{atm} - P_{abs}. \tag{6.3}$$

Many unit operations in food processing are done either at pressures above or below atmospheric pressure. The absolute pressure in equations (6.2) and (6.3) is often referred to as the operating pressure (P_{op}). Measurement of the P_{gauge} makes it possible to calculate the operating pressure.

Worked Problem 6.1

If the operating pressure in a retort reads 16.2 psi, express the retort pressure in units of (a) psia (lb_f/in^2), (b) atmospheres (atm), and (c) pascals (Pa).

Solution:

1. *Motivation*: One must understand how to convert between gauge, atmospheric, and absolute pressures. Different calculations require pressure to be expressed in different units.
2. *Define the problem*: We know the gauge pressure is 16.2 psi and are asked to determine the absolute pressure and express this value in three different unit systems. We know that the pressure value given is the gauge pressure, because the problem states that the pressure is in the retort. We know that the absolute pressure is required because of the units of the answer. "psia" implies absolute pressure.
3. *Think about the problem*: The absolute pressure is the sum of the atmospheric and gauge pressures. The atmospheric pressure, for most problems, may be considered constant at 14.7 psi, so the absolute pressure should be easily calculable. The rest of the problem consists of converting from one unit system to another.
4. *Plan the solution*: The absolute pressure will be calculated from the gauge pressure, and then conversions will be made to the other units.
5. *Implement the solution strategy*:
    a. Absolute pressure can be calculated from equation (6.2),

$$P_{abs} = P_{gauge} + P_{atm}.$$

Knowing the gauge and atmospheric pressures, we can calculate the absolute pressure,

$$P_{abs} = P_{gauge} + P_{atm} = 16.2 \ psi + 14.7 \ psi = 30.9 \ psia.$$

b. Now, the absolute pressure can be converted to units of atmospheres. The conversion factor from Appendix 1 is

$$1 \ atm = 14.7 \ psia. \tag{6.4}$$

So,

$$P_{abs} \times conv. \ factor = 30.9 \ psia \times \frac{1 \ atm}{14.7 \ psia} = 2.1 \ atm.$$

c. Likewise, the conversion factor for psia to Pa from Appendix 1 is

$$1 \ psia = 6.895 \times 10^3 \ Pa = 6.895 \ kPa \tag{6.5}$$

$$P_{abs} \times conv. \ factor = 30.9 \ psia \times \frac{6895 \ Pa}{psia} = 2.13 \times 10^5 \ Pa.$$

6. *Check the solution*: No explicit check of this solution exists. However, the values seem reasonable. A gauge pressure of 16.2 psi is a little above atmospheric pressure (14.7 psi or 101.325 kPa).
7. *Generalize the solution strategy*: Many problems require one to differentiate between gauge and absolute pressure. Conversion of units is an important aspect in solving food engineering problems. Many systems of units are used, despite recent efforts to consistently use SI units. In the U.S. food industry, the English system of units is most common, and the food scientist must know how to convert from SI to English units and vice versa.

Practice Problem 6.2

In freeze drying, it is necessary to pull a vacuum, where absolute pressure may be as low as 50 μm Hg. Express this value as a vacuum in mm Hg.

6.2   Gas laws

Except under extreme conditions of temperature and pressure, gases follow the ideal gas law,

$$PV = nRT. \tag{6.6}$$

The pressure (P) is an absolute pressure. The parameter V is volume, generally that of the container the gas is in; for example, the headspace inside a can. The parameter n refers to the number of moles of gas. Avogadro's number (N_A) often is required to calculate n. Strictly speaking, there are 6.02×10^{23} molecules in 1 g mol, and there are 10^3 g mol in a kg mol. If the mole prefix is omitted, it is likely that g mol are meant. In working a problem, if the mass (m) and molecular mass (M) of a gas (or any other molecule) are known, the magnitude of n is given by the ratio of mass to molecular mass ($n = m/M$). The parameter T is the absolute temperature. A common mistake when using the ideal gas law is to fail to express T on the absolute scale. If the temperature is Celsius (°C), the absolute temperature is on the Kelvin (K) scale,

$$K = °C + 273.16. \tag{6.7}$$

When doing simple textbook calculations, 273.16 is commonly rounded off to 273. If temperature is Fahrenheit (°F), the absolute temperature is on the Rankin (°R) scale:

$$°R = °F + 460 \tag{6.8}$$

The parameter R in equation (6.6) is the universal gas constant. Its magnitude can be obtained by rearranging the ideal gas equation and solving for R.

Table 6.1 Universal Gas Constant (R) in Various Units

Magnitude	Units
82.057×10^{-3}	$(L \cdot atm)/(g \, mol \cdot K)$
82.057	$(cm^3 \cdot atm)/g \, mol \cdot K$
8314	$J/(kg \, mol \cdot K)$ or $(m^3 \cdot Pa)/(kg \, mol \cdot K)$
8.314	$J/(g \, mol \cdot K)$
8.314×10^7	$erg/(g \, mol \cdot K)$
1545	$(ft \cdot lb_f)/(lb \, mol \cdot {}^{\circ}R)$

One g mol of any ideal gas occupies 22.4 L at 0°C and 1 atm pressure. Thus,

$$R = \frac{1 \text{ atm} \times 22.4 \text{ L}}{1 \text{ g mol} \times 273.16 \text{ K}} = 0.082 \frac{L \cdot atm}{g \, mol \cdot K}. \tag{6.9}$$

The magnitude of R in a variety of other units is listed in Table 6.1.

A couple of limiting cases are instructive and useful when considering the ideal gas law. For constant temperature (isothermal) processes, the product of pressure and volume is constant:

$$PV = \text{constant} \tag{6.10}$$

This is known as Boyle's law. It means that if P and V change at constant T, we can say

$$P_1 V_1 = P_2 V_2 \tag{6.11}$$

where the subscripts 1 and 2 refer to the two conditions of P and V.

A second limiting case is Charles's law. At constant volume, the ratio of P to T is constant:

$$\frac{P}{T} = \text{constant} \tag{6.12}$$

and

$$\frac{P_1}{T_1} = \frac{P_2}{T_2}. \tag{6.13}$$

Charles's law predicts a linear relationship between P and T at constant V. In terms of the ideal gas law, we have

$$P = \frac{nR}{V}T. \tag{6.14}$$

The slope of a graph of P vs. T is equal to nR/V following this expression. Since one mole of an ideal gas occupies 22.4 L at 1 atm of pressure and 0°C, a

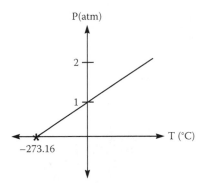

Figure 6.2 Determination of absolute zero

graph of P in units of atm vs. T in units of °C can be expressed as

$$P = 1 \text{ atm} + \frac{n\text{R}}{V} T \qquad (6.15)$$

$$= 1 \text{ atm} + \frac{1 \text{ mole} \times 0.082 \text{ (L} \cdot \text{atm})/(\text{mole} \cdot \text{deg})}{22.4\text{L}} \cdot T$$

$$= 1 \text{ atm} + (0.003660714 \text{ atm}/\text{deg}) \cdot T.$$

We can now pose the question: What temperature does this expression predict when pressure is zero? As shown in Figure 6.2, if P on the left side of equation (6.15) is set equal to 0, the temperature becomes equal to −273.16°C, which is called the absolute zero temperature.

Worked Problem 6.3

Oxygen gas occupies a volume of 10 L with a pressure of 102 kPa and a temperature of 25°C. Calculate the number of moles of O_2 present.

Solution:

1. *Motivation*: Gases and vapors are frequently encountered in food processing systems. Some food products exist in the gaseous state during processing steps, and some gases (such as drying air) are used in processing steps to further refine a product. The ideal gas law may be applied to most vapor systems to allow one to calculate properties of the gas.
2. *Define the problem*: We are given the volume, pressure, and temperature of the oxygen in this problem and asked to determine how many moles of oxygen are in the vessel.
3. *Think about the problem*: The ideal gas law relates each of these parameters with the gas law constant, R. We need to choose R in the correct units.

4. *Plan the solution*: The ideal gas law, equation (6.6), will be rearranged to solve for the number of moles. In this problem, we will use R = 0.082 (L·atm)/(g mol·K) and convert units of pressure from Pa to atm. This conversion is 1 atm = 101,325 Pa.

5. *Implement the solution strategy*: Rearranging the ideal gas law, equation (6.6), and solving for n gives:

$$n = \frac{PV}{RT} = \frac{(102,000 \text{ Pa})(10 \text{ L})}{\left(0.082 \frac{\text{L·atm}}{\text{g mol·K}}\right)(298 \text{ K})}\left(\frac{1 \text{ atm}}{101,325 \text{ Pa}}\right) = 0.4 \text{ moles}$$

6. *Check the solution*: We know that any gas at standard temperature (0°C) and pressure (1 atm) will occupy 22.4 liters for each mole. While we do not have exactly standard conditions in this problem, they are close. Our volume is a little less than half of 22.4 L, so the number of moles in 10 L should be a little less than 0.5 moles. Our answer seems reasonable.

7. *Generalize the solution strategy*: One of the keys to ideal gas law problems is to use the correct gas law constant. There are many forms to this constant, depending on the units of the system, and the correct value must be used. Using the ideal gas law, one can calculate volume, pressure, temperature, or the number of moles in the system, as long as three of the parameters are known.

Practice Problem 6.4

Air has about 21% O_2 and 0.03% CO_2. A room at 1 atm and 25°C has CO_2 pumped in until the pressure is 2 atm. Calculate the mole fraction of O_2 at the end of the process (R = 0.082 [L·atm]/[g mol·K]).

6.3 Gas mixtures

6.3.1 Partial pressure

Since different gases behave identically in regard to the ideal gas law, it follows that total pressure (P_T) of a mixture of gases is equal to the sum of the partial pressures (p_i) of the individual gases in that mixture:

$$P_T = \sum p_i \tag{6.16}$$

where i refers to the ith gas. Because of this, the following is true:

$$p_i V_T = n_i RT \tag{6.17}$$

$$P_T V_T = n_T RT \tag{6.18}$$

and

$$\frac{p_i}{P_T} = \frac{n_i}{n_T} = x_i = \text{mole fraction of gas } i. \tag{6.19}$$

This is Dalton's law of partial pressures. Equation (6.19) is useful for calculating mole fractions of gases from known partial pressures, or for calculating partial pressures from known mole fractions.

The gaseous state of water poses a special problem in food systems. Whereas gases like N_2, O_2, and CO_2 are considered permanent gases (they have boiling points far below room temperature), water vapor is not a permanent gas (its boiling point is close to room temperature). Fortunately, water in the vapor state follows ideal gas behavior. When working with air–water vapor mixtures, it is common to use the following expression:

$$P_T = p_{air} + p_w \tag{6.20}$$

where the first term on the right refers to the permanent gases in the air (e.g., N_2, O_2, CO_2, Ar), which are also known as dry air, and the second term is the partial pressure of water vapor. It is important to remember here that the ideal gas law fails to account for condensation of gases. Thus, if water vapor condenses during a particular process, the use of the ideal gas law (and consequently equation [6.20]) can lead to erroneous calculations.

From the standpoint of shelf-life stability of foods, water activity (a_w) is a very useful parameter. Water activity is normally approximately equal to the relative vapor pressure (RVP),

$$RVP = \frac{p_w}{p_w^o} \approx a_w \tag{6.21}$$

where $-p_w^o$ is the saturation vapor pressure (the partial pressure of water vapor at equilibrium associated with pure water at the same temperature where p_w is determined). Saturation vapor pressure is dependent on temperature only, and values can be found on a steam table such as those available online at the Engineering Toolbox http://www.engineeringtoolbox.com and in food engineering textbooks or engineering handbooks.

6.3.2   Partial volume

Dalton's law is based on partial pressures. Amagat's law is based on partial volumes (v_i) and can be handled analogously to Dalton's law:

$$P_T v_i = n_i RT \tag{6.22}$$

$$P_T V_T = n_T RT \tag{6.23}$$

where

$$V_T = \sum v_i \tag{6.24}$$

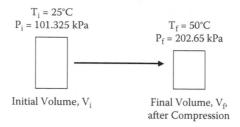

$T_i = 25°C$
$P_i = 101.325 \text{ kPa}$

$T_f = 50°C$
$P_f = 202.65 \text{ kPa}$

Initial Volume, V_i

Final Volume, V_f, after Compression

Figure 6.3 Effect of compression on moist air properties in worked problem 6.5

and therefore

$$\frac{v_i}{V_T} = \frac{n_i}{n_T} = x_i = \text{mole fraction gas } i. \tag{6.25}$$

Worked Problem 6.5

Saturated air is compressed at room temperature (25°C) and atmospheric pressure (101.325 kPa) until the total pressure is doubled and the temperature is 50°C, as shown in Figure 6.3. What is the final partial pressure of the water vapor? Note that at saturation, $p_w^o (25°C) = 2.52 \text{ kPa}$ and $p_w^o (50°C) = 12.3 \text{ kPa}$.

Solution:

1. *Motivation*: Many processing steps contain gases that are mixtures of more than one component, as in modified-atmosphere packaging, for example. The total pressure in the system is the sum of the individual partial pressures of the components. An understanding of these principles is necessary to determine how the properties of these gas mixtures change in the system.
2. *Define the problem*: The total pressure at both temperatures is known. We are asked to determine the final partial pressure of the water in the system as the temperature goes from 25°C to 50°C.
3. *Think about the problem*: Since the air is saturated at the start, the initial partial pressure of the water is equal to the equilibrium vapor pressure at the initial temperature. As the air is compressed, water may or may not condense. Initially, the water will be assumed to stay in the vapor phase. The ideal gas law may be used to determine how the air properties change during compression and temperature change. The gas volume changes, but the number of moles remains the same.
4. *Plan the solution*: The ideal gas law equation will be rearranged to solve for the number of moles of moist air in the system at the initial and final states. Since the number of moles is constant, initial and final conditions can be equated. From here, a ratio of the volumes will be

determined. This will be repeated for the water vapor. These ratios will be set equal to one another, and the partial pressure of the water vapor will be calculated.

5. *Implement the solution strategy*: Rearranging the ideal gas law, equation (6.6), to solve for the total moles of moist air gives

$$n = \frac{PV}{RT} = \frac{P_i V_i}{RT_i} = \frac{P_f V_f}{RT_f}.$$ (6.26)

Rearranging equation (6.26) to develop a ratio of the initial and final volumes gives

$$\frac{V_i}{V_f} = \frac{P_f T_i}{P_i T_f} = \frac{(202.65 \text{ kPa})(25°C + 273 \text{ K})}{(101.325 \text{ kPa})(50°C + 273 \text{ K})}$$

$$= \frac{(202.65 \text{ kPa})(298 \text{ K})}{(101.325 \text{ kPa})(323 \text{ K})} = 1.845.$$

This same equation may be written for the water vapor,

$$\frac{V_i}{V_f} = \frac{p_{w_f} T_i}{p_{w_i}^o T_f} = \frac{(p_{w_i})(298 \text{ K})}{(2.52 \text{ kPa})(323 \text{ K})} = 1.845.$$

Rearranging and solving for the vapor pressure gives

$$p_{w_f} = \frac{V_i}{V_f} \frac{p_{w_i}^o T_f}{T_i} = \frac{(1.845)(2.52 \text{ kPa})(323 \text{ K})}{(298 \text{ K})} = 5.04 \text{ kPa}.$$

This is lower than the vapor pressure of water at this temperature (12.3 kPa), so the assumption that the water does not condense is valid, and the partial pressure is 5.04 kPa. However, the air is no longer saturated with water vapor.

6. *Check the solution*: We have already performed one check by making sure the "no condensation" assumption was followed.

7. *Generalize the solution strategy*: The ideal gas law was used to relate the total pressure and volume of the air to the partial pressure and volume of the water vapor. An understanding of the ideal gas law allows one to calculate changes in headspace conditions, changes in vapor composition during process changes, and requirements for modified or controlled atmospheric storage.

Practice Problem 6.6

Calculate the pressure inside the headspace of a can at 15°C if the product was at 82°C at fill and the atmospheric pressure was 101.3 kPa at the time of sealing. Assume that the number of moles of air trapped in the headspace is constant and that the air is saturated. At saturation, p_w^o (15°C) = 1.7 kPa and p_w^o (82°C) = 51.3 kPa.

chapter seven

Mass balances

According to the law of conservation of matter, mass cannot be destroyed; it can only be converted from one form to another. This principle allows us to account for all the material that enters and exits a food processing operation or even the entire plant. The mass of incoming materials must be accounted for in either the product exiting the plant or in waste streams. Furthermore, the mass of individual components (water, protein, fat, etc.) must also be balanced within the process. That is, all of the protein (or sugar, water, etc.) must be balanced between the inputs and outputs. In fact, any individual specific component (any molecular species) must be balanced in a unit operation.

Mass balances are not only useful for tracking various ingredients and components within a food processing facility. They are also useful for product formulation, optimizing efficiency of a unit operation, and providing needed information for ingredient labeling, among other applications.

A mass balance accounts for the material entering and exiting a process, as well as whatever mass accumulates (or is lost) within the process. Furthermore, the mass of certain components may be generated or used up (consumed) in the form of a chemical reaction, and this mass must also be accounted for. The general equation that applies to all mass balances is

Accumulation of mass within the system

= sum of inputs − sum of outputs   (7.1)

+ generation − consumption

The solution to equation (7.1) depends on the circumstances under which the mass balances are to be used. The first step in any mass problem is to define exactly what problem must be solved. Usually, this requires drawing a flow diagram or figure that describes the process. For example, a simple mass-balance problem involves mixing two different materials together to get a product with a different composition. The composition of the final product depends on the characteristics of the initial materials and the ratio of their addition. A flow diagram for such a process would have two inputs (feeds) and one output (product), as seen in Figure 7.1.

The next step in solving mass balances is to draw a control surface around the desired operation. In our example, the operation is the mixing point, and we can draw a control surface around it, as shown in Figure 7.2.

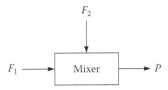

Figure 7.1 Schematic of process

In more complex food processing operations (multistage process, recycle streams, etc.), there may be numerous choices for drawing control boundaries, and part of the skill in solving mass-balance problems arises from knowing which control surfaces will lead to simple solutions.

Once all of the flow streams have been shown on the flow diagram, all known values (total mass, concentrations, etc.) should be labeled so that it is clear what is known and what needs to be calculated. Once the flow diagram is complete and the control surfaces are drawn, then solving a mass balance comes down to selecting the correct control surface on which to start writing the mass balances. In our simple example, there is only one control surface, so our mass balance involves total-balance (all mass) and component-balance (on individual components) equations. Equation (7.1) is applied for each material, and the resulting equations are solved together to give the unknowns and solve the mass balance.

In general, the number of equations that can be written around a boundary is equal to the number of components, n, plus 1. That is, a balance can be written on each nth component and on the total amount of mass for a total of $(n + 1)$ equations. However, only n of these equations are independent, meaning they represent new and unique equations. A dependent equation may be defined as one that can be derived from other independent equations.

At times, solving mass balances can be simplified by looking beyond individual components. For example, in a system that contains three components dissolved in water, we would have three independent equations. Instead of writing balance equations for each component, however, we might find that writing a balance on total dissolved solids (the sum of the three different components, or the nonwater components) allows us to solve the mass-balance equations more readily.

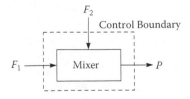

Figure 7.2 Schematic of process showing control surface

In some problems, it is possible that insufficient information is available to complete the mass balance. If all the equations are written and still no solution can be obtained, it usually means there is insufficient data. In this case, one or more of the parameters becomes a variable parameter, meaning that its value can be chosen by the processor. In some problems where ratios are used, the problem solver can choose to select a basis for the calculation. Either the mass of materials in or the mass of product out can be used to define the basis for ratio or percentage calculations. For example, sometimes it is necessary to specify, say, 100 kg of feed input or 100 lb of product out, so that all other parameters are given on a percentage basis.

In some mass balances, certain relationships among variables may be specified based on process requirements. Parameters like solubility of sugars or salts, specific percentage conversion of reactions, desired separation efficiencies, or specific ratios of ingredients may be required. For example, the process for making jam requires that the ratios of fruit, sugar, and pectin be specified. These ratios are specific process requirements that provide an additional equation to help solve mass-balance problems.

In the following sections, some examples of mass balances under different conditions are shown. We provide details and example problems for the case where we have steady-state operation and no reaction, steady-state operation with chemical reaction, and finally, unsteady-state operation (without reaction). These examples provide a good introduction to mass balances and the skills necessary to solve them.

7.1 Steady-state operation, no reaction

When no chemical reaction takes place, the last two terms in equation (7.1) (generation and consumption) are zero. Furthermore, for steady-state systems, where the mass does not accumulate within the system, the only terms remaining in equation (7.1) are the inputs and outputs. Thus, the mass-balance equation states simply that the mass of material that enters the process must equal the mass of material that exits. That is,

$$\text{sum of inputs = sum of outputs.} \tag{7.2}$$

Assuming a simple system with one input (F) and two outputs (P and V), the mass balance can be written as

$$F = P + V. \tag{7.3}$$

However, we can also balance the components within each product stream. Assuming we have a binary system, with two components (water and solids), then we can write mass balances on each of the individual components too, since water is not converted to solids and vice versa. The component balances are

$$\text{water in = water out} \tag{7.4}$$

$$\text{solids in = solids out.} \tag{7.5}$$

Since the total mass in each stream is the sum of water and solids, the three equations (7.3, 7.4, and 7.5) are not independent. That is, we can add two of the equations to obtain the third. For example, if equations (7.4) and (7.5) are added, we get equation (7.3), and thus not all the equations are independent. Only two of them are independent. This means we can only obtain useful information by solving two of the three equations. The third (dependent) equation will not give additional information.

To find the amount of an individual component present in a flow stream, the mass fraction is multiplied times the total amount of mass in that stream. For example, if an input stream entering a process contains 80% water and 20% solids, and a total of 100 lb are input, the amount of water entering the process is 0.80 (80%) times 100 lb, or 80 lb. In the same way, the amount of solids entering the process in that stream would be 0.20 (20%) times 100 lb, or 20 lb. Note that the total mass is the sum of the two individual components.

With these simple equations, numerous problems in food processing can be solved. The following problems (7.1 through 7.8) demonstrate the use of these equations in food processing calculations.

Worked problem 7.1

A moist food (100 lb) containing 80% water is dried to 50% moisture. How much water was removed?

Solution:

1. *Motivation*: This is a common type of problem in the food industry. Drying is often used to preserve foods, and the amount of moisture to be removed is an important parameter governing the size of the dryer.
2. *Define the problem*: We are asked to calculate the amount of water to be removed during the drying process. This will require a mathematical solution. A drawing will help visualize how these components are changing during the process (Figure 7.3). The food will be labeled F, and the product will be labeled P. We know that we have 100 lb of F, the moisture content in F is 80%, and the moisture content in P is 50%. We also assume that we remove pure water (100%), which contains no volatile compounds.

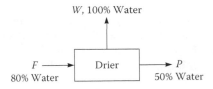

Figure 7.3 Schematic of drier

3. *Think about the problem*: Since this is a mass-balance problem, conservation-of-mass principles should be used. In our system, the amount of water must be conserved, as well as the amount of solids in the food.

4. *Plan the solution*: We must track the total mass, mass of water, and mass of solids in the system. There are two unknowns in this problem: W, the mass of water, and P, the mass of the food product. Therefore, two equations will be necessary to find the solution. These equations will come from the mass balances. The problem can be solved most easily if we use the total mass balance and the solids mass balance. We chose the solids mass balance because there are no solids in the water being removed; thus, the associated equation will be slightly easier to solve.

5. *Implement the solution strategy*: Writing a mass balance on the total amount of material in the system gives

mass of moist food = mass of water leaving

+ mass of final food product

$$100 \text{ lb} = W + P. \tag{7.6}$$

A mass balance on the amount of solids gives

mass of solids in moist food = mass of solids in drying water

+ mass of solids in final product

$$(0.20) = 0 + P(0.50) \tag{7.7}$$

$$(100 \text{ lb})(0.20) = 0 + P(0.50)$$

$$20 \text{ lb} = 0 + 0.5P = 0.5P.$$

Solving equation (7.7) gives

$$P = 20 \text{ lb}/(0.5) = 40 \text{ lb}.$$

Substituting $P = 40$ lb into equation (7.6) and solving for W gives

$$W = 100 - P = 100 - 40 = 60 \text{ lb}.$$

Thus, 60 lb of water will be removed in the process.

6. *Check the solution*: Are these numbers reasonable? They seem reasonable, since they are both less than 100 lb and clearly add up to 100 lb. We can check the answer by performing a mass balance on water. Note that this is the third, or dependent, equation, but it still must give a correct solution.

mass of water in moist food = mass of water removed

+ mass of water in final product

$$(0.80)F = W + (0.5)P$$

$$(0.80)\,(100\ \text{lb}) = 60\ \text{lb} + (0.50)\,(40\ \text{lb}) \tag{7.8}$$

$$80\ \text{lb} = 60\ \text{lb} + 20\ \text{lb} = 80\ \text{lb}.$$

The answer checks correctly.

7. *Generalize the solution strategy*: This strategy not only produced the amount of water removed, but also the amount of food product exiting the drier. So, this strategy would be applicable to that problem also.

Practice problem 7.2

A reverse-osmosis membrane unit concentrates a liquid food from 10% to 20% total solids by selectively removing water. If 100 lb/h of feed is input to the membrane, how much concentrated product is produced?

Worked problem 7.3

An unsalted product flowing at 50 kg/h is salted by pumping in salt solution (25%). How much salt must be added to make a product containing 1% salt? What is the product flow rate after salt addition?

Solution:

1. *Motivation*: This problem is quite common in food processing operations. Many food processes use a pumping system to mix ingredients. In this case, an ingredient is added in a concentrated form.
2. *Define the problem*: We are given the mass flow rate of the unsalted product (U), as well as the information that the product contains no salt. The salt percentages of the brine (S) and final product (P) are given, but the mass flow rates are unknown. Mass balances will be necessary to calculate these flow rates (Figure 7.4).
3. *Think about the problem*: Again, conservation-of-mass principles will be used. The total mass flow, salt, and other mass in the system must be conserved.
4. *Plan the solution*: For two unknowns, two equations must be used. The first equation will be generated by the total mass balance. The second

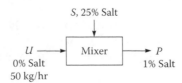

Figure 7.4 Schematic of mixer

equation will come from the salt balance or balance of the nonsalt mass. We will choose the salt balance because it will be slightly easier overall to solve. Since the unsalted product does not contain any salt, the salt balance will then only have two components, whereas the balance on the nonsalt mass will have three terms (check this yourself).

5. *Implementation of strategy*: Writing a mass balance on the total mass flow in the system gives

mass flow of unsalted food + mass flow of brine

$$= \text{mass flow of salted food product}$$

$$50 \text{ kg/h} + S = P. \tag{7.9}$$

A mass balance on the flow of salt gives

mass flow of salt in feed + mass flow of salt in brine

$$= \text{mass flow of salt in final product}$$

$$(0)U + (0.25)S = (.01)P \tag{7.10}$$

$$(0.25)S = (0.01)P. \tag{7.11}$$

Solving for P as a function of S,

$$P = (0.25)S/(0.01) = 25S. \tag{7.12}$$

Substituting equation (7.12) into equation (7.9) gives

$$50 \text{ kg/h} + S = 25S. \tag{7.13}$$

Solving for S,

$$24S = 50 \text{ kg/h}$$

$$S = 50 \text{ kg/h}/24 = 2.08 \text{ kg/h}. \tag{7.14}$$

Substituting this into equation (7.10) to solve for P,

$$50 \text{ kg/h} + 2.08 \text{ kg/h} = P = 52.08 \text{ kg/h}. \tag{7.15}$$

Thus, 2.08 kg/h of concentrated salt is needed to give 1% salt content in the mixed product.

6. *Check the solution*: Are these numbers reasonable? They seem reasonable, since the solution requires just 2.08 kg/h of brine to be added

to 50 kg/h of feed to produce a 1% salt solution. We can check the answer by performing a mass balance on the nonsalt mass.

$$\frac{\text{mass flow of}}{\text{unsalted food}} + \frac{\text{mass flow of nonsalt}}{\text{mass in brine added}} = \frac{\text{mass flow of nonsalt}}{\text{mass in final product}}$$

$$50 \text{ kg/h} + (1 - 0.25)S = (1 - 0.01)P$$

$$50 \text{ kg/h} + (0.75)S = (0.99)P \qquad\qquad (7.16)$$

$$50 \text{ kg/h} + (0.75)(2.08 \text{ kg/h}) = 51.56 \text{ kg/h}$$

and

$$(0.99)(52.08 \text{ kg/h}) = 51.56 \text{ kg/h}.$$

The answer checks correctly.

7. *Generalize the solution strategy*: This solution strategy produces the answer in relatively simple steps. Both flow rates were calculated. Any mixing process can be solved in a similar manner.

Practice problem 7.4

Assume that 30 kg of one component with 30% total solids (TS) is mixed with 200 kg of a material with 80% TS. Calculate the amount and composition of the final mixture.

Worked problem 7.5

How much juice flavor concentrate (25% TS) should be added to 100 lb of apple juice (9% TS) to produce a product with 12.5% TS? How much of this product will be produced?

Solution:

1. *Motivation*: This problem might occur in a recycle loop or situation where a product close to the final product in formulation is added to a feed solution during the process. In fact, many juice products are made by adding a concentrated juice to a less expensive juice to create a certain flavor.

2. *Define the problem*: We are asked to calculate the amount of concentrate to be added to a juice product to produce a juice with a different solids content. We are also asked to calculate how much product will be formed. Figure 7.5 will help visualize how these components are changing during the process. The feed juice will be labeled F; the concentrate will be labeled C; and the product will be labeled P. We know that we have 100 lb of F, that the solids content in F is 9%, and that the solids content in P is 12.5%.

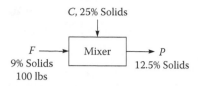

Figure 7.5 Schematic of mixer

3. *Think about the problem*: This is a mass-balance problem because conservation of mass must apply. In this case, both the amount of water and the amount of solids in the juice must be conserved.
4. *Plan the solution*: We must track the total mass, mass of water, and mass of solids in the system. There are two unknowns in this problem: C, the mass of water, and P, the mass of the juice product. Therefore, two equations will be necessary to find the solution. These equations will come from the mass-balance equations. The total mass balance will generate one equation, and the other will come from the solids-content balance. The water balance is the third equation, which in this case is dependent and can only be used for checking the answer.
5. *Implement the solution strategy*: Writing a mass balance on the total amount of material in the system gives

mass of feed juice + mass of concentrate

= mass of juice product

$$100 \text{ lb} + C = P. \tag{7.17}$$

A mass balance on the amount of solids gives

$$\frac{\text{mass of solids}}{\text{unsalted food}} + \frac{\text{mass of solids}}{\text{in concentrate}} = \frac{\text{mass of solids}}{\text{in juice product}}$$

$$F(0.09) + C(0.25) = P(0.125) \tag{7.18}$$

$$(100 \text{ lb})(0.09) + C(0.25) = P(0.125) \tag{7.19}$$

$$9 \text{ lb} + C(0.25) = P(0.125). \tag{7.20}$$

Dividing both sides by 0.125 gives

$$72 \text{ lb} + 2C = P. \tag{7.21}$$

Substituting equation (7.17) into equation (7.21) to eliminate P gives

$$72 \text{ lb} + 2C = 100 \text{ lb} + C. \tag{7.22}$$

Solving equation (7.22) gives

$$C = 28 \text{ lb}.$$

Substituting $C = 28$ lb into equation (7.17) and solving for P gives

$$P = 100 \text{ lb} + C = 100 \text{ lb} + 28 \text{ lb} = 128 \text{ lb}.$$

Thus, 128 lb of juice is produced.

6. *Check the solution*: Are these numbers reasonable? We can check the answer by performing a mass balance on water.

$$\frac{\text{mass of water}}{\text{in feed juice}} + \frac{\text{mass of water}}{\text{in concentrate}} = \frac{\text{mass of water in}}{\text{final juice product}}$$

$$(0.91)F + (0.75)C = (0.875)P \tag{7.23}$$

$$(0.91)100 \text{ lb} + (0.75)28 \text{ lb} = 91 \text{ lb} + 21 \text{ lb} = 112 \text{ lb}$$

and

$$(0.875)128 \text{ lb} = 112 \text{ lb}.$$

The answer checks correctly.

7. *Generalize the solution strategy*: This strategy not only produced a value for the amount of concentrate to add, but also produced the total mass of juice. The strategy confirms the value of knowing how to solve two equations for two unknowns. What would you expect if you mixed three or four or even five different ingredients? How might this problem be turned around? What would you do if you knew how much product you wanted with a desired concentration and were asked to find out how much of each ingredient you needed?

Practice problem 7.6

Add 55 lb of a raw material containing 15% alcohol and 1% sugar (the rest is water) to 45 lb of another raw material containing 17% alcohol and 0.5% sugar. What is the composition of the mixed product?

Mass balances can also be written on systems where a phase change occurs but the new phase is not separated from the product, as in freezing of foods or solidification of fats. Even if no change in mass occurs in a process, conservation equations can be written to distinguish the different phases of material.

For example, during freezing, liquid water is partially converted to ice, and the liquid phase composition changes according to how much ice was frozen. An example of this type of problem is given below where ice and liquid water can be separated to allow the freeze concentration of the liquid phase during freezing to be calculated.

Worked problem 7.7

Ice cream mix containing 40% total solids is frozen so that half of the original water is made into ice (which is effectively pure water). If the mix contained

12% fat, 16% sugar, and 12% nonfat milk solids, what is the concentration of fat, sugar, and milk solids in the aqueous phase of the frozen product?

Solution:

1. *Motivation*: This is a common type of problem in ice cream processing and any freezing process. During freezing, the remaining material becomes concentrated and changes characteristics.
2. *Define the problem*: We are asked to calculate how the composition of the liquid portion of the ice cream mix changes when half of the water freezes. A drawing will help visualize how these components are changing during the process. The mix (F) and ice cream (IC) are shown in Figure 7.6. The fat, sugar, and nonfat milk solids are known before freezing to be 12%, 16%, and 12%, respectively. Ice cream can be treated as a mixture of ice and concentrate. All of the fat, sugar, and nonfat milk solids are concentrated into the concentrate.
3. *Think about the problem*: In our system, the amount of water must be conserved, as well as the amount of solids in the ice cream. Since we know that half of the water in F freezes (this is an example of a process requirement and is actually based on the freezing point depression curve of the ice cream mix), we can determine how much ice we have coming out of the freezer. We can use mass-balance equations to determine the composition of the concentrate.
4. *Plan the solution*: We must track the total mass, mass of water, and mass of solids in the system. Since no mass enters or leaves, the mass of the ice cream is the same as the mass of the feed mix. The mass of the components does not change. However, the mass of water (liquid) changes due to the freezing. Mass balances on each component must be performed to determine the new composition in the liquid portion. Since no masses are given, an assumption (or basis) of total mass of 100 lb will be used.
5. *Implement the solution strategy*: Given 100 lb of mix, the amount of water in the mix is 100(0.6) = 60 lb. Since half of it freezes, 30 lb of

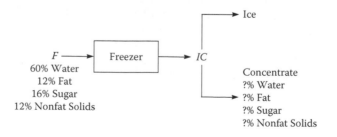

Figure 7.6 Schematic of freezer

water is frozen into ice, and a mass balance on the water in its various forms (liquid and ice) can be written as

mass of water in the mix = mass of water remaining

+ mass of water frozen in the concentrate

60 lb = water in concentrate + 30 lb

water in concentrate = 60 lb − 30 lb = 30 lb.   (7.24)

Thus, the total mass of the unfrozen liquid decreases from 100 lb initially to 70 lb (100 − 30) in the concentrate.
– Analysis of fat composition:

mass of fat in the mix = 100(0.12) = 12 lb

The 12 lb of fat does not change. The composition of fat in the concentrate is

$$X_F = 12 \text{ lb}/70 \text{ lb} = 0.171.$$

– Analysis of sugar composition:

mass of sugar in the mix = 100(0.16) = 16 lb

The 16 lb of sugar does not change. The composition of sugar in the concentrate is

$$X_S = 16 \text{ lb}/70 \text{ lb} = 0.229.$$

– Analysis of nonfat solids composition:

mass of nonfat solids in the mix = 100(0.12) = 12 lb

The 12 lb of nonfat solids (NFS) does not change. The composition of nonfat solids in the concentrate is

$$X_{NFS} = 12 \text{ lb}/70 \text{ lb} = 0.171.$$

6. *Check the solution*: Are these numbers reasonable? The compositions should increase, since the liquid mix is basically concentrated in this process. We can check the water composition in the concentrate. From our analysis of the other three components, the water composition should be 0.429 (1 − 0.171 − 0.229 − 0.171).

mass of water in the mix = 100(0.30) = 30 lb

The composition of water in the concentrate is

$$X_W = 30 \text{ lb}/70 \text{ lb} = 0.429.$$

The answer checks correctly.

7. *Generalize the solution strategy:* This problem can be deceiving. One cannot just assume that the compositions will double, since half of the water is removed by freezing. The conservation-of-mass principles must be used carefully. How much water must be removed as ice to cause the solution concentration to double?

Practice problem 7.8

Fondant is a highly crystallized sugar matrix used in the confectionery industry. A sugar solution containing 12% water with dissolved solids at a ratio of 80:20 sucrose to corn syrup solids is crystallized so that 51.2% of the fondant product is crystalline sucrose. Assuming that all of the corn syrup solids remain dissolved, calculate the water content, total dissolved solids, corn syrup solids content, and dissolved sucrose content of the liquid phase of this fondant.

More complex problems can be solved using the same principles as applied in the previous example problems. Multicomponent problems with more than a single unit operation can be solved by application of the same principles, although the optimal choice of control boundaries to solve around the equations and the order of solving the equations can be significantly more complicated.

7.2 Steady-state operation, with chemical reaction

In some situations, chemical reactions take place within the unit operation that involve consumption of certain components and generation of others. When no reaction occurs, mass and moles are always equivalent and both balance. However, when a reaction occurs, the stoichiometry of the reaction governs the decrease in substrate concentration(s) and increase in product concentration(s). Thus, moles of each species must be "balanced" according to the stoichiometry of the reaction; however, conservation of mass must still hold.

For example, a simple, irreversible chemical reaction where component A reacts to form components B and C can be described by the following equation:

$$A \rightarrow B + C \tag{7.25}$$

Such an equation, for example, might describe the hydrolysis of sucrose to glucose and fructose. In this case, one mole of sucrose is converted into one mole each of fructose and glucose. The proper mass balance is in terms of moles of reactants and products. Note that the total number of moles entering a process may not be the same as the number of moles exiting. It is not truly moles that are conserved, except stoichiometrically, but mass that must be balanced. Converting from mass to moles (and back again) requires use of the molecular weight of each reactant or product.

Worked Problem 7.9

A solution containing 10 moles of sucrose is partially (25%) inverted to fructose and glucose during a hold at elevated temperature. Calculate the number of moles of each at the end of this process.

Solution:

1. *Motivation*: In many situations in the food industry, sugar-based products may be held at elevated temperature, leading to substantial inversion (hydrolysis) of sucrose. This reaction can have significant impact on product characteristics, since fructose and glucose have different properties (crystallization, sweetness, boiling point elevation, glass transition, etc.) than sucrose.
2. *Define the problem*: We initially start with 10 moles of sucrose (S) and water (W). Some of the sucrose (25% of the initial level) is converted to glucose (G) and fructose (F) during the process. If we make a diagram, we will have 10 moles of sucrose entering our process, and unknown values of sucrose, glucose, and fructose exiting the process, as seen in Figure 7.7.
3. *Think about the problem*: This is a mass-balance problem where a chemical reaction takes place that consumes some of the reactant (sucrose) and generates new products. At the end of the process, we have 75% of the initial sucrose present and unknown levels of fructose and glucose. The reaction can be written as

$$S + W \rightarrow F + G. \tag{7.26}$$

   One mole each of sucrose and water is needed for inversion of sucrose to the reaction products, glucose and fructose.
4. *Plan the solution*: We will use a mass balance based on the stoichiometric ratio of moles of reactants and products to calculate the unknown concentrations.
5. *Implement the solution strategy*: A balance on moles of sucrose is

   Moles into process = 10.

   Exiting the process, 75% of the initial moles of sucrose are remaining:

   Moles out of process = 10 (0.75) = 7.5

Figure 7.7 Schematic of reaction

A balance on glucose is

Moles into process = 0.

Since for each mole of sucrose hydrolyzed, one mole of fructose is produced,

Moles out of process = 10 (0.25) = 2.5.

A similar calculation applies for glucose. So, exiting the process, we have 7.5 moles of sucrose and 2.5 moles each of fructose and glucose. On a molar basis, the total number of moles out is 12.5, even though only 10 moles of sucrose entered. This is not creation of mass, since the molecular weight of sucrose (342) is higher than that of fructose and glucose (180), but we can check the mass balance to verify that our calculation is correct.

6. *Check the solution*: The best way to check that this solution is correct is to verify that mass has indeed been conserved. We can calculate the mass of reactants and products by multiplying the number of moles by the molecular weight.

Mass of sucrose entering = 10 moles (342 g/mole) = 3420 g sucrose

Mass of water used = 2.5 moles (18 g/mole) = 45 g water

Mass of sucrose exiting = 7.5 moles (342 g/mole) = 2565 g sucrose

Mass of fructose exiting = 2.5 moles (180 g/mole) = 450 g fructose

Mass of glucose exiting = 2.5 moles (180 g/mole) = 450 g glucose

The total mass entering the process, sucrose (3420 g) plus water (45 g) = 3465 g, should equal the total mass exiting the process. The sum of sucrose, glucose, and fructose exiting the process is 2565 + 450 + 450 = 3465 g, the same mass as entering the reaction scheme. Thus, the masses balance in this problem.

7. *Generalize the solution strategy*: In any process involving a chemical reaction, mass must be balanced, but so must the number of moles of reactants and products according the stoichiometry of the reaction. Suppose in the chemical reaction of *A*, two moles of product *B* and one mole of product *C* were produced. How would this change the calculation?

Practice problem 7.10

Natural gas is fed into a furnace and burned such that 90% of the limiting reactant is consumed. Natural gas contains 20 mole% CH_4, 60 mole% O_2, and 20 mole% CO_2. The reaction that describes combustion of natural gas is

$$CH_4 + 2O_2 \rightarrow CO_2 + 2H_2O.$$

Find the molar composition of the exit flue gas.

7.3   Unsteady-state operation, no reaction

In the case where mass is either entering a system faster than it leaves or leaving faster than it enters, there will be an accumulation (or loss) within the system that is time dependent. For example, if water is filling a bucket with a hole in the bottom, the bucket either becomes more full or less full, depending on whether the flow in is faster or slower than the flow out. The change in level of water in the bucket over time can be determined based on unsteady-state mass-transfer calculations. Such a problem is very important in the food processing industry, where product streams are often pumped into holding tanks prior to introduction to the next unit operation in the process. Unsteady-state mass balances can also be used to determine how long it will take to fill or empty a vessel given a certain pumping rate.

Equation (7.1) becomes a differential equation, since the rate of change in mass with time (dm/dt) within the system, the accumulation term, is balanced by the input and output terms.

$$(dm/dt) = \text{sum of inputs} - \text{sum of outputs} \tag{7.27}$$

To solve such unsteady-state problems, the differential equation that arises from equation (7.27) must be solved with the appropriate boundary conditions. Examples of problems solved with the unsteady-state equation are given in problems 7.11 and 7.12.

Worked problem 7.11

A tank containing 1000 lb of milk is being pumped out until the tank is empty. If the pump removes milk at a rate of 5 lb/min, how long will it take to empty the tank?

Solution:

1. *Motivation*: Any situation where a vessel is being drained can be solved with unsteady-state mass balances. This process might occur in an ice cream plant preparing for a new batch, or a cheese plant unloading fresh milk.
2. *Define the problem*: We are asked to calculate the time required to empty the tank. This is an unsteady-state mass-balance problem, where conservation-of-mass principles must be written in terms of their change with time. We are given the pump flow rate and the initial mass of the milk in the tank, as seen in Figure 7.8.
3. *Think about the problem*: This is an unsteady-state problem, and our approach will be slightly different than for previous problems.

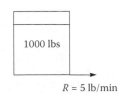

Figure 7.8 Milk removal from tank

4. *Plan the solution*: Look at the general equation (7.27) for unsteady-state problems. That equation will determine how the solution proceeds. In this case, the rate of change in mass within the tank must balance the rate of product removal.

5. *Implement the solution strategy*: We begin with the general unsteady-state equation,

$$\text{mass in} - \text{mass out} = \text{mass accumulated.} \tag{7.28}$$

There is no mass in. The mass accumulated (or lost, in this case) is actually the mass leaving the tank at the rate of milk being pumped out.

$$-\text{out} = \text{accum} = \frac{dm}{dt} \tag{7.29}$$

So, for 5 lb/min as the rate of milk out of the tank,

$$\frac{dm}{dt} = -5\frac{\text{lb}}{\text{min}}. \tag{7.30}$$

To solve this differential equation, we need to separate variables and integrate over the proper boundary conditions. Multiplying both sides of equation (7.30) by dt gives

$$-dm = 5dt. \tag{7.31}$$

Now, integrating both sides with a closed integral from the initial condition of $t = 0$ and $m = 1000$ lb to the final condition where $t = t$ and $m = 0$, when all the milk has been removed from the tank,

$$-\int_{1000}^{0} dm = 5\int_{0}^{t} dt \tag{7.32}$$

$$-m\big|_{1000}^{0} = 5t\big|_{0}^{t} \tag{7.33}$$

$$-(0-1000) = 1000 = 5(t-0) = \left(5\frac{\text{lb}}{\text{min}}\right)t. \tag{7.34}$$

Solving for t,

$$t = 1000 \text{ lb}/(5 \text{ lb/min}) = 200 \text{ min.}$$

6. *Check the solution*: This answer may be checked intuitively by dividing the 1000-lb mass by the rate of 5 lb/min to obtain 200 min for the removal time. The formal mass balance in this case can be solved easily by this intuitive method. In more complex problems, the solution will not be so intuitive.

7. *Generalize the solution strategy*: There are many other applications where the liquid level in a tank changes. The rate of change in level depends on the relative rates of product flowing in and out. How would you go from rate of change of mass of material to rate of change in level? Remember that mass equals density times volume ($m = \rho V$) and that volume can be expressed as area times height (or level).

Practice problem 7.12

A retention tank (10 m³) stores fruit juice ($\rho = 1050$ kg/m³) as it flows between two processing steps. Fruit juice is continually fed into the tank and, ideally, is removed at the same rate to maintain constant volume. During a process upset, the outflow stream is decreased from 10 kg/min to 5 kg/min. If the inflow stream remains at 10 kg/min and the tank was initially half full, how long will it take the tank to overflow?

chapter eight

Energy balances

According to the second law of thermodynamics, energy cannot be created or destroyed, just changed from one form to another. For example, frictional energy is turned into heat, and potential energy is turned into kinetic energy, but the total energy must be conserved. In many food systems, we can deal with a simplification of a complete energy balance and simply account for enthalpy changes of a material as it moves through a unit process. In this case, energy inputs (e.g., mechanical energy, kinetic energy, and potential energy) are not important, and we can simply compare the internal energy of a material within a system. These other energy terms will be discussed further in the fluid flow section, where we perform a mechanical energy balance and assume that changes in internal energy are not important.

The enthalpy of a material is a relative measure of its heat content and is a combination of two terms: sensible heat and latent heat. Sensible heat is a measure of the energy change of a material above a certain reference temperature, T_{ref}, and is given as

$$\Delta H = mc_p \Delta T \tag{8.1}$$

where
$\quad \Delta H$ = enthalpy change (J)
$\quad m$ = mass (kg)
$\quad c_p$ = heat capacity (J/kg·K)
$\quad \Delta T$ = temperature difference from T_{ref}.

The heat capacity of a material, c_p, is a measure of how much energy can be stored in the material, or the amount of energy required to increase the temperature of 1 lb of material by 1 °F. Heat capacity of water is defined as 1 Btu/lb·°F (4.2 kJ/kg·K) at 100 °C.

Note that enthalpy is not an absolute value, but is relative to the reference temperature chosen. This reference temperature can be any value, as long as that value is used consistently throughout the calculation. Oftentimes, 0 °C (32 °F) is used as T_{ref} because the steam tables are based on this temperature.

The second component term of enthalpy is latent heat, which applies if a material undergoes a phase change. The energy associated with that phase change, either taken in or released, also contributes to enthalpy. Enthalpy change associated with latent heat can be calculated as

$$\Delta H = m\lambda.$$

(8.2)

where m is the mass (kg or lb) of material that actually undergoes the phase change, and λ is the latent heat (J/kg or Btu/lb).

For example, if 1 lb of water is heated from 32 °F to its boiling point (212 °F), the enthalpy change is calculated as sensible heat from equation (8.1), using 32 °F as our reference temperature:

$$\Delta H = (1 \text{ lb}) (1 \text{ Btu/lb} \cdot {}^{\circ}\text{F}) (212 - 32 \,{}^{\circ}\text{F}) = 180 \text{ Btu}$$

or, in SI units,

$$\Delta H = (1 \text{ kg}) (4.2 \text{ kJ/kg} \,{}^{\circ}\text{C}) (100 - 0 \,{}^{\circ}\text{C}) = 420 \text{ kJ}.$$

If that 1 lb of water is heated further, it vaporizes and turns into steam, increasing enthalpy through addition of latent heat. The enthalpy change of heating 1 lb of water to 212 °F, and then boiling off the last drop of water so that we have saturated steam at 212 °F, is calculated by summing the sensible heat change with the latent heat change. For this calculation we need to know that the latent heat of vaporization of water at 1 atmosphere of pressure is 973 Btu/lb (2257.1 kJ/kg). That is, it takes 973 Btu to convert 1 lb of water at 212 °F to 1 lb of steam at 212 °F. Thus, the enthalpy change for the entire process is

$$\Delta H = mC_p \Delta T + m\lambda$$

$$\Delta H = (1 \text{ lb}) (1 \text{ Btu/lb} \cdot {}^{\circ}\text{F}) (212 - 32 \,{}^{\circ}\text{F}) + (1 \text{ lb}) (970.3 \text{ Btu/lb}) \qquad (8.3)$$

$$\Delta H = 1150.3 \text{ Btu}$$

or, in SI units,

$$\Delta H = (1 \text{ kg}) (4.2 \text{ kJ/kg} \,{}^{\circ}\text{C}) (100 - 0 \,{}^{\circ}\text{C}) + (1 \text{ kg}) (2257.1 \text{ kJ/kg})$$

$$\Delta H = 2677.1 \text{ kJ/kg}.$$

Substantially more energy is required to cause the water to vaporize, and this is reflected in the substantially larger enthalpy of the steam than the water at boiling point. This difference in enthalpy between steam and water at the boiling point is very useful in food processing, as condensation of steam to release this huge latent heat is used to provide an efficient heating source. Many food processes use condensing steam to heat the food.

Enthalpy balances are extremely useful tools in the food processing plant. They can be used to design and size equipment for heat exchange; they can be used for energy audits and conservation and for determining the amount of energy required in a processing step; and, in conjunction with mass balances, they are used to design certain unit operations (evaporation, drying, distillation, etc.).

8.1   Steam tables

One of the most important tools used for enthalpy balances in the food industry is the "steam table." This is a compilation of various properties of water in both liquid form and vapor, or steam. Specifically, the steam table provides the temperature at which water boils at any pressure, the enthalpy (relative to 0 °C or 32 °F) of water and steam, and the enthalpy required to go from water to steam (or vice versa). Steam tables are divided into "saturated steam" and "superheated steam" tables.

8.1.1   Saturated steam

Properties of saturated steam, which is defined as water vapor at its equilibrium pressure and temperature, may be found from either the pressure or temperature of the steam. If the pressure of the water–steam system is increased, then the vaporization (and condensation) temperature also increases. On the other hand, if a system operates under a vacuum (pressure lower than atmospheric), then the vaporization (and condensation) temperature is decreased (compared with the boiling point at atmospheric pressure). This relationship between equilibrium temperature and pressure is given in the steam tables. In many food processes, operation at pressures other than atmospheric is important to product quality. For example, many food evaporators are operated under vacuum to reduce the boiling temperature and minimize thermal changes in the food.

Since evaporation and condensation of water are important processes in the food industry, an understanding of steam tables and how to use them to calculate enthalpy requirements in a unit operation is needed. Steam tables provide information on enthalpy of pure water under various conditions. From the steam table, one can obtain enthalpy values, always using 32 °F (or 0 °C) as the reference temperature, for water at any temperature and pressure, and steam at any temperature and pressure. Thus, enthalpy changes for pure water can be determined simply by taking the difference between the appropriate entries within the steam table for the conditions pertinent to the process. For example, the enthalpy change required to heat water from 32 °F to saturated steam at 212 °F is found by subtracting the enthalpy values for these two states, as found in the steam table. For steam at 212 °F and 1 atm pressure, the enthalpy, H, is 1150.3 Btu/lb, and for water at 32 °F, the enthalpy is 0 Btu/lb (enthalpy here is zero because 32 °F is our reference temperature). The change in enthalpy, ΔH, is just the difference between these two, or 1150.3 Btu, as determined earlier (see equation [8.3]).

An understanding of the steam tables is critical to solving enthalpy balances in the food industry. The following problems are intended to help the reader gain familiarity with the steam tables and their use.

Worked problem 8.1

A fluid food is being evaporated at a pressure of 7.5 psia. What is the product's temperature as it evaporates?

Solution:

1. *Motivation*: It is important to understand what a food product's temperature is during any evaporation step. To maintain product quality, evaporators are operated under vacuum to keep boiling temperature low. This minimizes product degradation.
2. *Define the problem*: We know that the food is at a pressure of 7.5 psia and are asked to determine the temperature of the water as it evaporates. The steam tables must be used for the solution, since the steam table provides the equilibrium relationship between pressure and temperature.
3. *Think about the problem*: This problem requires us to find the boiling temperature of water at the given pressure. Water evaporates at the boiling temperature of water at the specified pressure (assuming no boiling-point elevation). The steam table provides this relationship between boiling temperature and pressure.
4. *Plan the solution*: The steam table (Table 8.1) has entries at pressures of 5 and 10 psia, but not at 7.5 psia. The answer must be obtained by interpolating between the table entries.
5. *Implement the solution strategy*: Table 8.1, a section from the steam table, will be used for this problem.
   The linear interpolation solution strategy will be employed.
   a. Table 8.1 contains entries that surround 7.5 psia.
   b. The entries that surround 7.5 psia are 5 and 10 psia. The boiling temperatures at these two points are 162.24 °F (T_5) and 193.21 °F (T_{10}), respectively.
   c. Finally, compute the fractional difference.

Table 8.1 Selected Entries from Saturated-Steam Table

Pressure [psia]	Temperature °[F]	H_f [Btu/lb]	H_g [Btu/lb]	H_{fg} [Btu/lb]
5	162.24	130.20	1,131.1	1,000.9
10	193.21	161.26	1,143.3	982.1
20	227.96	196.27	1,156.3	960.1
25	240.07	207.59	1,160.2	952.65

$$FD = (\text{unknown pressure} - \text{low pressure})/$$
$$(\text{high pressure} - \text{low pressure})$$
$$FD = (7.5 \text{ psia} - 5 \text{ psia})/(10 \text{ psia} - 5 \text{ psia})$$
$$= 2.5 \text{ psia}/5 \text{ psia} = 0.5$$

Now, the solution is

$$T_{7.5} = FD \ (T_{10} - T_5] + T_5$$
$$T_{7.5} = 0.5(193.21 \text{ °F} - 162.24 \text{ °F}) + 162.24 \text{ °F}$$
$$T_{7.5} = 0.5(30.97 \text{ °F}) + 162.24 \text{ °F} = 15.49 \text{ °F} + 162.24 \text{ °F} = 177.73 \text{ °F}$$

The boiling temperature at 7.5 psia is 177.73 °F.

6. *Check the solution*: In this case, we have assumed that the change in temperature is linear with the change in pressure (linear interpolation). Thus, the boiling temperature at 7.5 psia should be halfway between the boiling temperatures at 5 and 10 psia. That is,

$$T_{7.5} = (162.24 + 193.21)/2 = 177.72 \text{ °F}.$$

Thus, the answer checks (within rounding error).

7. *Generalize the solution strategy*: This is the basic interpolation technique necessary for many steam-table problems. Anytime data is presented in tabular form, interpolation may be needed to find the values between table entries.

Practice problem 8.2

What is the pressure of steam vaporizing at 150.5 °F?

Worked problem 8.3

Ten pounds of saturated steam at 20 psia condenses (condensate at saturation temperature) in the jacket of a steam-heated kettle.

1. What is the steam temperature?
2. How much energy is produced by this process?
3. This energy is used to heat 100 lb of a food ($c_p = 0.92$ Btu/lb- °F). How large a temperature increase will the food experience?

Solution:

1. *Motivation*: This is a classic problem in the food industry, where a product is being heated in a steam kettle. The steam tables can be used

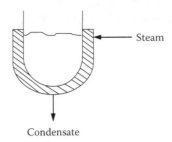

Steam

Condensate

Figure 8.1 Schematic of steam kettle

to determine how much steam is needed to heat a product to a certain temperature, or how hot it will get if a certain amount of steam is condensed.

2. *Define the problem*: We know the conditions of the steam (10 lb at 20 psia) and the conditions of the food (100 lb with a specific heat of 0.92 Btu/lb-°F). This is a simple energy balance problem (Figure 8.1) that will require the law of conservation of energy (energy from steam goes into heating the food). We will also need to determine properties of steam from the steam table.

3. *Think about the problem*: Part (1) of this problem requires us to find the condensing temperature at the given pressure. Assuming no heat losses, the heat released by the steam as it condenses will be absorbed by the food to raise the product temperature.

4. *Plan the solution*:
   a. Use the steam table (Table 8.1) to find the condensing temperature.
   b. The enthalpy released during complete condensation (H_{fg}), which also can be found in the steam table, is used to calculate the energy produced.
   c. Use the specific heat and mass of the food to determine the increase in temperature for the food. The energy released upon condensation of the steam goes into heating up the food.

$$q = \Delta H_{steam} = \Delta H_{food} \tag{8.4}$$

5. *Implement the solution strategy*:
   a. Table 8.1 will be used to solve part (a) of step 4 of this solution. At a pressure of 20 psia, the condensation temperature is 227.96 °F. No interpolation is needed.
   b. Also from Table 8.1, the H_{fg} is 960.1 Btu/lb. This is the amount of energy given up by the steam as it condenses. For 10 lb of steam, the heat released is

$$q = mH_{fg} = (10\ \text{lb})\left(960.1\frac{\text{Btu}}{\text{lb}}\right) = 9601\ \text{Btu}.$$

c. We assume that the heat absorbed by the food, which is equal to the heat released by the steam, follows equation (8.2).

$$q = mc_p \Delta T \tag{8.5}$$

and from equation (8.4)

$$q = (100 \text{ lb})\left(0.92 \frac{\text{Btu}}{\text{lb F}}\right) \Delta T = 9601 \text{ Btu.} \tag{8.6}$$

Rearranging,

$$\Delta T = \frac{q}{mc_p} = \frac{9601}{(100)(0.92)} \frac{\text{Btu}}{\text{Btu}/^\circ \text{F}} = 104.4^\circ \text{ F.} \tag{8.7}$$

So, the temperature increase in the food is 104.4 °F.

6. *Check the solution*: For this particular food product, the condensation of 10 lb of steam raises the temperature of 100 lb of the food by 104.4 °F. This is a large increase, but not unreasonable. In this case, there is no quantitative method of verifying the accuracy of this calculation.

7. *Generalize the solution strategy*: This problem illustrates the concept of conservation of energy and is valuable for the food scientist or engineer to understand. How can this problem be turned around? Suppose we know how much we wanted to heat our product, and asked how much condensed steam would be required.

Here is another idea. In this problem, we assumed that all of the heat from the steam condensing went into heating up the food. But in practice, heat leaks often cause the process to be inefficient, as some of the heat is lost into the surrounding environment. Thus, only a portion of the heat may be available to heat up the food. How could we modify the enthalpy balance for this case?

Practice problem 8.4

Five pounds of H_2O is heated from 70 °F to boiling at a pressure of 12 psia.

1. At what temperature does the water boil?
2. How much energy is required to heat the H_2O to boiling temperature?
3. How much energy is required to cause the H_2O to boil?

8.1.2   Superheated steam

The previous calculations were based on saturated steam, where temperature and pressure are related according to the equilibrium curve. Another aspect of the steam tables of importance to enthalpy balances in food processing occurs when steam is heated above the saturation temperature. That is, saturated steam heated above its equilibrium (condensation) temperature is called superheated steam. The degree of superheat is simply the difference in temperature above the saturation temperature. For example, if steam at one atmosphere (212 °F) is heated to 220 °F, it is superheated by 8 °F (220 – 212 °F). Superheated-steam tables can be used to find the enthalpy of steam at any temperature and pressure.

The following examples provide practice for using the superheated-steam tables.

Worked problem 8.5

Steam at 25 psia and 250 °F is condensed at its saturation temperature.

1. What is the temperature of the condensate?
2. How much energy is produced by this process?

Solution:

1. *Motivation*: It is common in food processing to find superheated steam as a source of heat. For example, steam may be heated above its equilibrium temperature in anticipation of heat losses during piping steam from the boiler to the application point in the processing plant.
2. *Define the problem.* We are given the steam's temperature (250 °F) and pressure (25 psia) and asked to find the temperature at which it will condense and how much energy will be released (Figure 8.2). The first task is to determine if this steam is saturated or superheated, and that will determine which steam table is needed.
3. *Think about the problem*: Part (1) of this problem requires us to find the condensing temperature at the given pressure. For this steam to be saturated, the boiling temperature at 25 psia should be 250 °F. If the

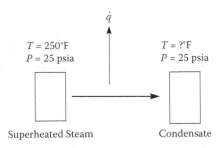

Figure 8.2 Heat loss in superheated steam

boiling temperature at this pressure is less than 250 °F, the steam is superheated by its difference from 250 °F. The energy produced by this process potentially includes two terms: the energy lost during cooling of superheated steam to the condensation temperature, and the energy released during condensation.

4. *Plan the solution*:
   a. Use the saturated-steam table (Table 8.1) to find the condensing temperature, and determine if this steam is superheated or not.
   b. The enthalpy of the steam can be found from the appropriate table. If the steam is superheated, we can calculate the sensible heat change as the superheated steam is cooled to its condensing temperature and then find H_{fg} from Table 8.1.
5. *Implement the solution strategy*:
   a. Table 8.1 will be used for part (a) of step 4 of this solution. At a pressure of 25 psia, the condensation temperature is 240.07 °F. This is less than 250 °F, so the steam is at a higher temperature than the equilibrium temperature, and is superheated by nearly 10 °C.
   b. Also from Table 8.1, the enthalpy of saturated steam at this pressure, H_g, is 1160.2 Btu/lb, and the enthalpy of water at this temperature, H_f, is 207.59 Btu/lb. From Table 8.2, the enthalpy of superheated steam at 250 °F and 25 psia (H_1) is 1165.6 Btu/lb.

The sensible heat, ΔH, as the superheated steam cools to its condensation point is given by the difference between the enthalpy of the superheated steam, H_1 (from Table 8.2), and the enthalpy of saturated steam, H_g (from Table 8.1).

$$\Delta H_s = H_1 - H_g \tag{8.8}$$

The latent heat given off by the steam condensing, H_{fg}, can be found as the difference between the enthalpy of saturated steam, H_g (from Table 8.1), and the enthalpy of water at that temperature, H_f (from Table 8.1).

$$\Delta H_{fg} = H_g - H_f \tag{8.9}$$

Table 8.2   Selected Entries from Superheated-Steam Table at $P = 25$ psia

Temperature [°F]	Enthalpy [Btu/lb]
250	1,165.6
300	1,190.2
350	1,214.5
400	1,238.5

Summing to get the total change in enthalpy from the superheated steam gives

$$\Delta H_{total} = \Delta H_s + \Delta H_{fg} = (H_1 - H_g) + (H_g - H_f) = H_1 - H_f. \qquad (8.10)$$

Solving for the heat released, using the values for H_1 and H_f given previously,

$$\Delta H_{total} = 1165.6\frac{Btu}{lb} - 207.59\frac{Btu}{lb} = 957.99\frac{Btu}{lb}.$$

6. *Check the solution*: We would expect the enthalpy for condensation of superheated steam to be a little higher than the enthalpy for condensation of saturated steam at the same pressure. From Table 8.1, H_{fg} is 952.65 Btu/lb at $P = 25$ psia, and the answer above is indeed slightly higher.

7. *Generalize the solution strategy*: Any situation where steam is at a temperature higher than its equilibrium boiling temperature requires the use of the superheated-steam tables, usually in combination with the saturated-steam tables. This problem also reinforces the concepts of sensible and latent heat, and how they are used together to determine the total heat produced in a system.

Practice problem 8.6

How much energy is needed to make superheated steam at 300 °C from saturated steam at 20 psia? Note: the pressure is constant.

8.2 Enthalpy balances

To solve enthalpy balances in food processing operations, a balance equation is needed that accounts for changes in enthalpy of a material as it goes through a process. In enthalpy balances, we need to balance all heat input terms as well as the enthalpies of the various flow streams that enter and exit a process operation. Furthermore, an enthalpy balance also accounts for changes associated with chemical reactions (exothermic or endothermic) and heats of solution. Thus, an enthalpy balance accounting for all these terms can be written as

$$\begin{pmatrix} \text{Accumulation of enthalpy} \\ \text{within a system} \end{pmatrix} = \begin{pmatrix} \text{sum of enthalpies} \\ \text{of input streams} \end{pmatrix} - \begin{pmatrix} \text{sum of enthalpies} \\ \text{of output streams} \end{pmatrix}$$

$$+ \begin{pmatrix} \text{heat} \\ \text{inputs} \end{pmatrix} - \begin{pmatrix} \text{heat} \\ \text{losses} \end{pmatrix} + \begin{pmatrix} \text{heat of} \\ \text{reactions} \end{pmatrix}. \qquad (8.11)$$

Under steady-state conditions, where enthalpy within a system does not change, the accumulation term is zero. Solving energy balances follows much the same process as we used for solving mass balance problems. A steady-state enthalpy balance involves an accounting of the enthalpy of the streams entering a process compared with those leaving the process. A flow diagram of the process can be drawn with input and output streams labeled. In addition to the flow rates, we also need to define temperature and phase (vapor, liquid, or solid) of our material to define the enthalpy terms of importance. The boundary of our system, whether a single operation or multiple processes, defines the region within which enthalpy must be balanced. A simple example is provided in worked problem 8.7, with a practice problem given in problem 8.8.

Worked problem 8.7

One hundred pounds of fluid food is heated by addition of 10,000 Btu of energy. If the food (C_p = 0.9 Btu/lb-°F) has an initial temperature of 40°F, what is the temperature after heating?

Solution:

1. *Motivation:* This problem illustrates a common practice in the food industry: heating a food with an outside heat source. In heat exchangers, the energy may come from steam condensing or a hot fluid flowing through the heat exchanger. When no phase change occurs in the food product, all of this heat energy goes into warming up the cool fluid.
2. *Define the problem:* We know the energy input into the system (10,000 Btu), and the specific heat (0.9 Btu/lb-°F) and initial temperature (40°F) of the food. We are asked to find the final temperature of the food product through an energy balance (Figure 8.3).
3. *Think about the problem:* This problem is very straightforward. The information given relates directly to the heat content equation, from which we can calculate change in temperature.
4. *Plan the solution:* The heat content equation will be used to directly solve for the output temperature. We know that as a fluid heats up, assuming no phase change occurs, the change in temperature can be calculated from

$$q = mc_p \, \Delta T. \tag{8.5}$$

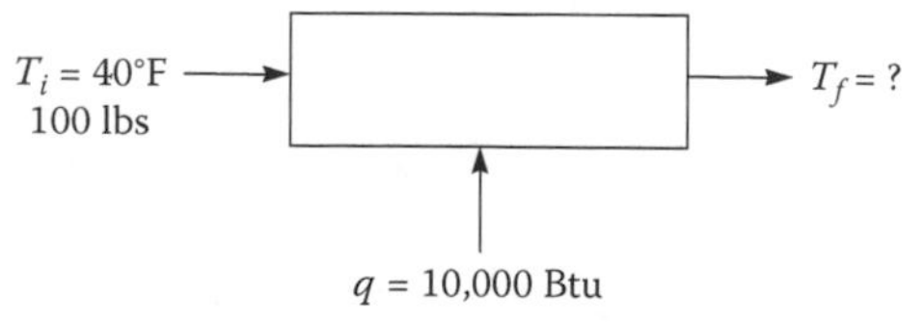

T_i = 40°F ⟶ ⟶ T_f = ?
100 lbs

q = 10,000 Btu

Figure 8.3 Schematic of heater

5. *Implement the solution strategy*: Expanding equation (8.5) to show initial and final temperatures, T_i and T_f, respectively,

$$q = mc_p(T_f - T_i). \qquad (8.12)$$

Rearranging to solve for T_f,

$$T_f = T_i + \frac{q}{mc_p}. \qquad (8.13)$$

Now, substituting the given values,

$$T_f = 40°\ F + \frac{10,000\ Btu}{(100\ lb)(0.9\ ^{Btu}/_{lb-°\ F})} = 40°\ F + 111.1°\ F = 151.1°\ F.$$

The temperature of the food after heating with 10,000 Btu of energy is 151.1 °F.

6. *Check the solution*: This seems like a reasonable temperature. It is higher than the initial temperature, as expected, but it is also less than the normal boiling point of water, so there is no danger of phase change associated with boiling.

7. *Generalize the solution strategy*: Here, we implemented the heat equation to calculate the temperature rise due to an outside heat source. We utilized a simple form of the enthalpy balance equation. A slightly more complicated problem might involve combining the calculations for enthalpy change in condensing steam (see worked problems 8.3 and 8.5), with the calculation for temperature change shown here.

Practice problem 8.8

If 4200 kg of a food ($c_p = 4.2$ kJ/kg-°C) loses the energy equivalent of 5000 kJ, then

1. How much does its temperature go down?
2. If there was initially only 420 kg of food, what would be the temperature loss?

Worked problem 8.9

Hot water ($c_p = 4.18$ kJ/kg-°C) initially at 90 °C is used to preheat juice ($c_p = 3.85$ kJ/kg-°C) from 4 °C to 50 °C prior to evaporation of the juice. If the flow rate of water is 200 kg/h and that of the juice is 150 kg/h, what is the outlet temperature of the water?

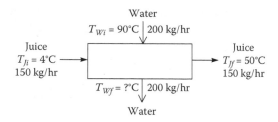

Figure 8.4 Heating of juice with hot water

Solution:

1. *Motivation:* Heating of a cold fluid by use of a hot fluid in a heat exchanger is extremely common in the food industry. From milk pasteurization to cooling of beer, numerous examples of this process can be found. It is very important to be able to calculate changes in temperature in such a system by employing an energy balance.

2. *Define the problem:* In this process, the hot water gives up some of its heat to the juice, so the water loses energy and the juice gains energy, as seen in Figure 8.4. Assuming no energy losses, the amount of energy lost by the water should balance the amount of energy gained by the juice.

3. *Think about the problem:* Equation (8.11) defines the energy balance for this system. We will need to sort through which of those terms are important in this problem, and then apply the energy balance to solve for the temperature of the water after it loses energy to the juice.

4. *Plan the solution:* Although the temperatures of both juice and hot water change, this is still a steady-state problem. The juice always comes out of the heat exchanger at the desired temperature, so even though the juice changes temperature within the heat exchanger, the temperature at the inlet and outlet are always constant. Thus, it is a steady-state process. In the same way, the temperature of the water exiting the heat exchanger will always be at a constant value as well.

   To solve this steady-state energy balance, all of the terms in equation (8.11) can be eliminated except the sum of input and output streams. That is,

$$\text{sum of input enthalpies} = \text{sum of output enthalpies.} \qquad (8.14)$$

The enthalpy of each stream can be defined by a form of equation (8.1) using flow rate $\dot{m}$,

$$\Delta H = \dot{m} c_p (T - T_{ref}) \qquad (8.15)$$

where the reference temperature, T_{ref}, can be defined as any temperature. Here, we will use $T_{ref} = 0\,°C$ to simplify the problem.

5. *Implement the solution strategy*: Writing the enthalpy terms for juice and water according to equation (8.14) gives

$$\Delta H_{Jin} + \Delta H_{Win} = \Delta H_{Jout} + \Delta H_{Wout} \tag{8.16}$$

$$m_J c_{pJ}(T_{Jin} - T_{ref}) + m_w c_{p,W}(T_{Win} - T_{ref}) = m_J c_{pJ}(T_{Jout} - T_{ref})$$
$$+ m_w c_{pW}(T_{Wout} - T_{ref}). \tag{8.17}$$

Assuming that c_p is the same for the inlet and outlet conditions (it is not a function of temperature), we can substitute the known parameters into equation (8.18).

$$(150 \text{ kg/h}) (3.85 \text{ kJ/kg-°C}) (4 - 0 \text{°C})$$
$$+ (200 \text{ kg/h}) (4.18 \text{ kJ/kg-°C}) (90 - 0 \text{°C})$$
$$= (150 \text{ kg/h}) (3.85 \text{ kJ/kg-°C}) (50 - 0 \text{°C})$$
$$+ (200 \text{ kg/h}) (4.18 \text{ kJ/kg-°C}) (T_{Wout} - 0 \text{°C})$$
$$2{,}310 + 75{,}240 = 28{,}875 + 836 \ (T_{Wout})$$
$$48{,}675 = 836 \ (T_{Wout})$$
$$T_{Wout} = 58.2 \text{°C}$$

The water temperature out of the heat exchanger is 58.2 °C.

6. *Check the solution*: Another approach to this problem, assuming no heat losses and that c_p is not a function of temperature, is to simply say that the enthalpy change of the juice has to equal the enthalpy change of the hot water.

$$\Delta H_J = \Delta H_W \tag{8.18}$$

$$\dot{m}_J c_{pJ}(T_{Jout} - T_{Jin}) = \dot{m}_w c_{pW}(T_{Win} - T_{Wout}) \tag{8.19}$$

$$(150 \text{ kg/h}) (3.85 \text{ kJ/kg-°C}) (50 - 4 \text{°C})$$
$$= (200 \text{ kg/h}) (4.18 \text{ kJ/kg-°C}) (90 \text{°C} - T_{Wout})$$
$$26{,}565 = (836) (90 - T_{Wout})$$
$$T_{Wout} = 58.2 \text{°C}$$

The answer checks in this approach.

7. *Generalize the solution strategy*: In equation (8.17), the choice of T_{ref} is arbitrary, and we often simply choose the most convenient temperature to simplify the calculations. In this particular example, because

there were no heat losses and C_p was constant, we could even go a step further to cancel out T_{ref} entirely. However, in many problems, this will not be the case, and you will need to keep T_{ref} in the enthalpy balance.

An extension of this type of problem is one where there is also a heat loss to the environment. It is normal for some of the energy in the hot water to be lost to the environment around the heat exchanger. This lost energy must also be accounted for in the enthalpy balance.

Practice problem 8.10

Hot water is used to heat milk (100 kg/h) from 2 °C to 43 °C in a heat exchanger. The hot water comes in at 90 °C and at a flow rate of 200 kg/h. The temperature of the water coming out of the heat exchanger has been measured to be 65 °C. How much of the energy in the hot water is lost to the environment through the walls of the heat exchanger?

---

chapter nine

Fluid mechanics

Fluids are described as materials that do not retain their shape, but take the shape of their container. Fluid foods include liquids such as water and oil, suspensions such as some fruit juices and purees, emulsions such as mayonnaise and ice cream, and dispersions such as those made with gums and starches along with the gels that form after they are heated. Fluid foods are also frequently heterogeneous mixtures of water, fat, proteins, and carbohydrates, which give them very complex behavior. The movement of these fluids must be controlled during unit operations such as pipe flow, pumping, pasteurization, mixing, filling, metering, etc. This provides the degree of processing and the inputs required by heat and mass transfer, as discussed in chapters 10 and 11. The principles used to predict how fluids will behave in a given system are referred to as fluid mechanics.

In this chapter, the topic of rheology, which is the study of flow and deformation of matter, is introduced first. The rheological properties of fluid foods are needed to determine how a fluid will react to an applied force. Then the fundamental laws of conservation of mass and energy, as well as a form of the conservation of momentum, are used to build the concepts needed to predict the way that material will flow in pipe systems and the amount of energy that needs to be added if flow is to occur at the desired rate. Finally, the fluid mechanics principles introduced in this chapter will be combined with an understanding of heat transfer in chapter 10, which will allow for the determination of heat-exchanger sizes or holding times needed to properly heat process fluid foods.

9.1 Rheology

In rheology, the nature of a material depends on its reaction to a force applied over a specific area. The force per unit area is the stress. When the force is applied perpendicular to the area, as is typically the case with the gravitational and pressure forces the resulting stress is called a normal stress. A force applied parallel to surface produces a shear stress, as demonstrated in Figure 9.1. When a shear stress is applied to a fluid, it continuously deforms, and the deformation is permanent. This situation is known as flow. Conversely, when a shear stress is applied to an elastic solid, it will deform in proportion to the force. However, it will immediately return to its original shape when the force is released. There is a third situation in which the

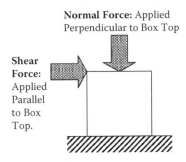

Figure 9.1 Forces applied to the top of a box; the bottom of the box is fixed to the ground

material will deform and only partially recover its original shape. Materials of this type are described as viscoelastic or plastic.

Many food materials display viscoelastic properties, including cheese, bread dough, cake batter, etc. However, for the purposes of this chapter, we will concentrate on fluid foods that do not display viscoelastic behavior.

In the case of simple fluids made up of small molecules, such as water and oil, Newton observed that the when the shear stress (σ) increased, the rate of deformation (also known as the strain rate or shear rate, $\dot{\gamma}$) increased proportionally. This means that, for the box with the fixed bottom shown in Figure 9.1, the speed (rate) that the top moves (deforms) in a block of fluid of a given height, increases in proportion to the force that is applied. The shear rate is defined in this simple case as the difference between the velocity of the top and the bottom of the box divided by the box height ($\dot{\gamma}$ = velocity difference/ height) to give a typical unit of 1/s. The relationship Newton observed is represented mathematically as

$$\sigma = \mu\dot{\gamma} \tag{9.1}$$

where the proportionality constant (μ) is called the Newtonian viscosity, with typical units found in Appendix 1.

In the case of more complex fluids, such as those containing biopolymers that include proteins, or long-chain carbohydrates that are typically used as food thickeners, the viscosity is no longer a constant. In this case, the relationship between the shear stress and the shear rate is described using a non-Newtonian rheological model.

9.1.1 Modeling rheological behavior of fluid foods

When a series of shear stress–shear rate data for an unknown fluid is plotted, it typically produces a curve like one of those found in Figure 9.2. In the case of a Newtonian fluid that follows the relation in equation (9.1), the data

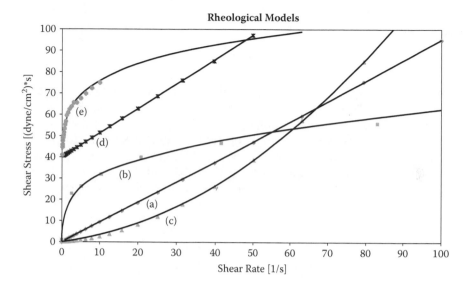

Figure 9.2 Common rheological behaviors seen in fluid foods that are predicted using (a) Newtonian rheological model, (b) shear-thinning power law, (c) shear-thickening power law, (d) Bingham plastic rheological model, and (e) Herschel–Bulkley rheological model

is in a straight line through the origin, as seen in curve (a). The slope of that line is the viscosity. If any measured shear stress is divided by the shear rate that generated it, the result will always give the same value for the viscosity. However, in all the other non-Newtonian cases shown, the ratio of the shear stress to the shear rate is dependent on the specific value of the shear rate, and is therefore known as the apparent viscosity:

$$\mu_a = \frac{\sigma}{\gamma}. \tag{9.2}$$

In the case of curve (b), the apparent viscosity goes down as the shear rate increases. This is known as shear thinning and is very common behavior in fluid foods. This behavior has been used to advantage in salad dressing, for example, since it allows us to easily shake spices into the dressing that are initially found at the bottom, due to the shear thinning caused by the shaking. The spices then remain suspended while we pour the dressing on our salad, and the dressing clings to the lettuce, due to the low shear rate provided by gravity. Curve (c) shows shear-thickening behavior, where the apparent viscosity increases with increasing shear rate. This type of behavior is rare in foods. The only commonly encountered example is corn starch suspended in water before heating.

Both curves (b) and (c) are cases of the rheological model known as the power-law model:

$$\sigma = k \cdot \gamma^n \qquad (9.3)$$

where k is the consistency coefficient with typical units of $Pa \cdot s^n$ or (dyne/cm^2)s^n, and n is the dimensionless flow behavior index. The flow behavior index is $n > 1$ for shear-thickening fluids, and $n < 1$ for shear-thinning fluids. When $n = 1$, the power-law expression reduces to the Newtonian case, and $k = \mu$. To determine the value of the power-law parameters from experimental data, the easiest way is to linearize equation (9.3) by taking the log or natural log of both sides of the equation, as described previously in section 3.3, to get in the natural log case:

$$\ln \sigma = \ln k + n \ln \dot{\gamma}. \qquad (9.4)$$

In this case, the parameters are found by curve fitting the natural logs of the shear-stress and shear-rate data to the equation of a line, where the slope will be the flow behavior index, n, and $e^{intercept}$ will be the consistency coefficient, k.

Curves (d) and (e) cross the y-axis at a stress value greater than zero. This stress value is known as the yield stress (σ_o), and represents the level of force that must be applied to a given area before flow will start. Yield stress is what you experience when the ketchup will not come out of the bottle unless you shake it. Shaking increases the force acting on the ketchup, so it begins to flow. Yield stress is highly desirable in foods like pudding or mashed potatoes, which need to keep their shape on a spoon but easily deform in the mouth. These two curves can be represented by the following expression:

$$\sigma = \sigma_o + k \cdot \dot{\gamma}^n. \qquad (9.5)$$

This expression is known as the Herschel–Bulkley model when $n < 1$, which corresponds to curve (e). When $n = 1$, which is the case in curve (d), it is known as the Bingham plastic model. If the yield stress is zero, this expression reduces to the power-law model. Therefore, this general expression can represent any of the curves in Figure 9.2, depending on the value of the parameters. The procedure for finding the parameters of the Bingham plastic model from experimental data was previously illustrated in section 3.2 in example 3.2.1. In addition, example 3.3.2 in section 3.3 shows how to linearize the experimental data and find the model parameters for the Herschel–Bulkley model when the value of the yield stress has been measured separately or estimated from the graph of shear stress versus shear rate.

One final important rheological model for food is the Casson model, which is the official model for modeling the flow behavior of melted chocolate,

$$\sigma^{0.5} = \sigma_o^{0.5} + k \cdot \dot{\gamma}^{0.5}. \qquad (9.6)$$

After transforming the data by taking the square root of the shear rate and shear stress, the model parameters (σ_o and k) can easily be found by fitting the transformed data to a straight line. Other rheological models, examples of fitting data to the models and typical model parameters for a wide range of food materials, have been compiled in books by Steffe (1996) and Rao (2007).

9.1.2   Measurement of rheological properties

The way that shear-stress and shear-rate data is obtained is to apply either a known force over a known area (stress) and measure the deformation to obtain the shear rate, or else to apply a known deformation over a known height and area of material and measure the force required to deform the material. In the case of fluid foods, the simplest approaches measure how much materials move in response to gravity, such as in dropping a ball into the fluid and seeing how long it takes to settle, or in measuring the time it takes for a fluid to run down an inclined plane. For example, in the falling-ball viscometer, the viscosity (or apparent viscosity) is measured by dropping a solid ball of a known density (ρ_s) into the center of a fluid (liquid) cylinder (radius = R) with a known density (ρ_l), and measuring the time (t) it takes to travel a given distance (L). Then the viscosity is found using the following equation:

$$\mu = \frac{2R^2 g t}{9L}(\rho_s - \rho_l) \tag{9.7}$$

where g is the gravitational constant.

While the simple force-of-gravity approaches are inexpensive and easy, they are prone to a lack of precision and only provide a single viscosity value. When it is necessary to examine fluids with subtle differences or that are non-Newtonian in nature, measurements are typically made on a viscometer, which provides a precise measurement of viscosity, or a rheometer, which provides a wide range of rheological data. The two most common types of instruments are rotational viscometers, which measure the torque or the angle of rotation generated by a rotating element, and capillary or tube rheometers, which measure the force or the pressure drop generated during flow in a pipe or capillary tube of a given length.

Rotational viscometers and rheometers come in a wide range of geometries. A rotating bob in a cup like that shown in Figure 9.3 is frequently used with thin liquids that would run off of a flat plate. For

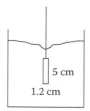

Figure 9.3 Rotational viscometer

5 cm

1.2 cm

Newtonian fluids, the viscosity can be determined from the relationship,

$$\mu = \frac{M}{8\pi^2 NL}\left(\frac{1}{R_i^2} - \frac{1}{R_o^2}\right) \tag{9.8}$$

where M is the torque, N is the rotational speed in revolutions per second, L is the length, R_i (inside radius) is the radius of the bob, and R_o (outside radius) is the radius of the cup. Variations of this design include replacing the bob with four-or six-blade vanes for use with materials containing particles or with high-yield stresses, and use of the bob in a large container, which is illustrated in the following problem.

Worked problem 9.1

In a wide-gap rotational viscometer, the cup radius (R_o) effectively goes to infinity, so the relationship between torque (M) and rotational speed (N) for a Newtonian fluid can be written as

$$M = 8\pi^2 \mu L R_i^2 N \tag{9.9}$$

where

μ = viscosity (Pa·s)
L = height of the spindle (m)
R_i = radius of the spindle (m).

For the data given below, calculate the viscosity of this fluid. A spindle with $L = 5$ cm and $R_i = 1.2$ cm was used. Remember to convert rpm to revolutions per second. Use a spread sheet and calculate the correlation coefficient to determine how well this data fits the equation.

N (rpm)	M (N·m)	N (rps)
0	0	0
5	7.50×10^{-6}	0.0833
10	1.71×10^{-5}	0.1667
15	2.30×10^{-5}	0.2500
20	3.02×10^{-5}	0.3333
25	3.75×10^{-5}	0.4167

Solution:

1. *Motivation*: An important parameter in formulation of a new product is the viscosity of the material. Viscosity influences both the texture of the final product and the flow properties during processing. Rotational viscometers are often used to find viscosity based on the relationship between torque and rpm.

2. *Define the problem:* We are given the torque and rotational speed data from a rotational viscometer and asked to determine the food's viscosity. Figure 9.3 shows the viscometer. Because this is a wide-gap rotational viscometer, the diameter of the vessel holding the fluid must be sufficiently large compared with the diameter of the rotating spindle.

3. *Think about the problem:* The rotational speed must be converted to revolutions per second so that the viscosity units will be correct. If M is plotted vs. N, a straight line should be found. According to equation (9.9), the slope (m) of the line will be equal to $8\pi^2 \mu L R_i^2$. These values are all constants, and the viscosity can be determined from this slope.

4. *Plan the solution:* After converting N to rps, M will be plotted vs. N in an Excel spreadsheet. If this is a straight line, the slope and the correlation coefficient will be calculated with a forced zero intercept. The viscosity will be calculated from the slope.

5. *Implement the solution strategy:* The spreadsheet solution is shown in Figure 9.4. First we converted rpm to rps by dividing the rpm values by 60 s/min. Then we plotted the points and fit them to a line with a zero intercept. The curve fit result and correlation coefficient (R^2) are shown on the graph. The R^2 value of 0.996 indicates an excellent fit of the data to the model, so we have a linear relationship with a zero intercept, as expected, and with a slope (m) of 9.142×10^{-5} N·m·s/rev. From equation (9.9), the slope is equal to $8\pi^2 \mu L R_i^2$. This equation may be rearranged and solved for viscosity.

$$\mu = \frac{m}{8\pi^2 L R_i^2} = \frac{9.142 \times 10^{-5} \frac{\text{N·m·s}}{\text{Rev}}}{8\pi^2 (0.05\ \text{m})(0.012\ \text{m})^2} = 0.161\ \text{Pa·s}$$

The viscosity of this Newtonian fluid is approximately 0.161 Pa·s or 16.1 cP.

6. *Check the solution:* We can verify the slope calculation by using a quick graphical approach to calculate the slope from two points on the line.

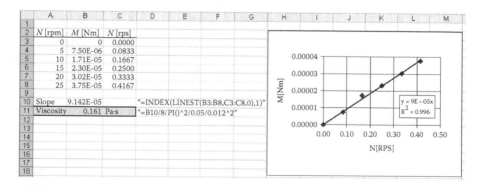

Figure 9.4 Excel solution to worked problem 9.1

The first and last data points seem to be on the line, and since we know their exact values, we will use them.

$$\text{slope} = m = \frac{y_2 - y_1}{x_2 - x_1} = \frac{(3.75 \times 10^{-5} - 0)\text{N} \cdot \text{m}}{(0.4166 - 0)\,\text{rps}} = 9.00 \times 10^{-5}\,\frac{\text{N} \cdot \text{m} \cdot \text{s}}{\text{rev}}$$

This is slightly different than the spreadsheet graphical approach, but it is close enough to verify the result and also provide an alternative approach if a spreadsheet program were unavailable.

7. *Generalize the solution strategy:* This problem required us to determine how to best use the data. Whenever a table of data is given, such as in this problem, a graph is often necessary. How confident would you feel calculating the viscosity from one data point?

For more viscous liquids, the rotational geometry of choice is narrow-gap parallel-plate or the small-angle cone-and-plate geometry. These narrow-gap geometries are extremely well understood and can be used for a wide range of rheological property measurements. While the torque and rotational velocity are always available, this type of instrument typically converts those values directly to shear-rate and shear-stress results. These results are then fit to a specific model, such as those described in section 9.1.1. The following practice problem shows an example of the data provided by these instruments and provides an opportunity to work through the process of obtaining the model parameters.

Practice problem 9.2

The apparent viscosity of a non-Newtonian food changes with shear rate ($\dot{\gamma}$) according to the following data. Find the consistency coefficient, $k[\text{Pa·s}^n]$, and the flow behavior index, n, for this power-law fluid.

$\dot{\gamma}(1/s)$	μ_a (Pa·s)
1	89.0
2	63.1
4	47.8
8	28.8
12	22.4
24	18.0

Capillary-tube viscometers are used to provide higher shear-rate ranges than can typically be generated in rotational viscometers. They work by establishing a pressure difference to create flow. The pressure force can be generated by gravity, as is the case for U-tube glass capillary viscometers for thin liquids such as milk, or by a piston or pump, as is the case for high-pressure capillary viscometers and pipe viscometers. Their operating principle is based on the principles for pipe flow that will be described in the next few sections.

9.2 Fluid flow

9.2.1 Continuity equation

As discussed in chapter 7 about mass balances, the principle of conservation of mass requires that as long as there is no material being stored or lost in a system, what goes in must come out. For the pipe in Figure 9.5, the area of the inlet (A_1) is different than the area of the outlet (A_2). In this flow situation, conservation of mass indicates that the mass flow rate in ($\dot{m}_1$) will equal the mass flow rate out ($\dot{m}_2$), where the mass flow rate is equal to the product of the volumetric flow rate and the density ($\dot{m} = \rho\dot{V}$). Then, using the fact that the volumetric flow rate is the product of the cross-sectional area and the average velocity ($\dot{V} = A\bar{v}$), this brings about the following relationship:

$$\rho_1 A_1 \bar{v}_1 = \rho_2 A_2 \bar{v}_2 \qquad (9.10)$$

which is the continuity equation for pipe flow with a single inlet and outlet. This relationship allows the determination of the effect of pipe diameter on the fluid average velocity, which is illustrated in the following two problems.

Worked problem 9.3

If a fluid food flowing at 1 m/s in a pipe with d_i = 0.02 m flows into a pipe of d_i = 0.04 m, what is the average velocity in the expanded pipe?

Solution:

1. *Motivation:* This is a problem common to the food industry. Many times, a piping system will expand or contract so that pipe diameters will meet fitting sizes on processing equipment. It is important for the food scientist or engineer to understand how these changes affect flow of the fluid.

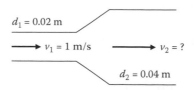

d_1 = 0.02 m

v_1 = 1 m/s v_2 = ?

d_2 = 0.04 m

Figure 9.5 Pipe expansion

2. *Define the problem*: The inlet pipe diameter and inlet velocity of the flowing food are given, as well as the outlet diameter. We are asked to find the outlet fluid velocity.

3. *Think about the problem*: This is a classic continuity problem. According to the continuity principle, the same amount (mass) of material must pass from point to point in time. Thus, the velocity must change as the pipe diameter changes.

4. *Plan the solution*: The solution will be found by using the continuity equation.

5. *Implement the solution strategy*: For constant density, the continuity equation (equation [9.10]) can be written as

$$\bar{v}_1 A_1 = \bar{v}_2 A_2 \tag{9.11}$$

where the area of a circle is

$$A_i = \frac{\pi}{4} d_i^2. \tag{9.12}$$

Therefore, after substituting and simplifying,

$$\bar{v}_1 d_1^2 = \bar{v}_2 d_2^2. \tag{9.13}$$

Rearranging equation (9.13) to solve for $\bar{v}_2$ and substituting in the known values,

$$\bar{v}_2 = \frac{\bar{v}_1 d_1^2}{d_2^2} = \frac{(1\,\tfrac{m}{s})(0.02\ m)^2}{(0.04\ m)^2} = 0.25 \frac{m}{s}.$$

The velocity in the expanded pipe is less than in the smaller pipe, as expected.

6. *Check the solution*: As stated above, the velocity in the larger pipe is less than that in the smaller pipe, as predicted by the continuity equation.

7. *Generalize the solution strategy*: This technique applies to both contractions and expansions in flow systems. Note that a two-fold increase in diameter results in a four-fold decrease in velocity according to the continuity equation.

Practice problem 9.4

What pipe diameter will give a mean flow velocity of 4 ft/min for a fluid being pumped at 2 gal/min? (Note: 1 gal = 0.13368 ft³.)

9.2.2 Determination of flow regime

When a faucet is turned on at a low flow rate, the stream coming out appears very smooth. If dye were continuously injected into the stream, it would come out as a straight line. This type of flow is found in what is called the laminar flow regime. Conversely, if the faucet is turned on at a very high flow rate, the flow of the stream coming out appears much more erratic, and the surface appears rough. If dye were injected into this stream, it would immediately begin to spread and mix into the entire flow. This type of flow is found in the turbulent flow regime. Between these two extremes at intermediate flow rates is the regime known as transitional flow. As we will see, the flow regime greatly affects the way fluids flow and the amount of energy that needs to be added by pumps in order to transport them. It also has a big effect on many of the other unit operations that occur in food processing, including those involving heat transfer (chapter 10) and mass transfer (chapter 11).

The flow regime is influenced by the fluid properties such as density and viscosity, the flow rate, and the dimensions of any fluid–solid interfaces that surround the flow. It depends on the balance between the forces of momentum and inertia that are being opposed by viscous and frictional forces. The dimensionless Reynolds number for a Newtonian fluid is

$$N_{Re} = \frac{\text{inertial forces}}{\text{viscous forces}} = \frac{\rho \bar{v} D}{\mu} \tag{9.14}$$

which provides a means to evaluate that balance. The variable D is the characteristic dimension of the flow system, which in the case of pipe flow is the inside diameter of the pipe. When the Reynolds number (N_{Re}) is low (<2100 for pipe flow), the flow regime is laminar. When the Reynolds number (N_{Re}) is high (>4000 for pipe flow), the flow regime is turbulent. The following two problems provide a means for gaining insight into the effects that the variables influencing the flow regime have on the Reynolds number (N_{Re}).

Worked problem 9.5

Milk is flowing at an average velocity of 0.459 m/s in a 1-in. nominal-size stainless steel sanitary pipe ($d_i = 0.02291$ m). For milk, $\rho = 1032$ kg/m^3 and $\mu = 1.33 \times 10^{-3}$ Pa·s. Is this turbulent flow or laminar flow?

Solution:

1. *Motivation*: Turbulent and laminar flow regimes are critical in determining what type and size of pump to select for pumping fluids. In addition, heat transfer and fluid mixing depend on the flow regime in flow systems.

2. *Define the problem*: The average velocity, density, viscosity, and pipe diameter for a flow system are given. We are asked to determine if this is laminar or turbulent flow.
3. *Think about the problem*: The problem appears straightforward, since the four components for the Reynolds number equation are given. Turbulence is based on the value of the Reynolds number. A Newtonian fluid (such as milk) is turbulent if the Reynolds number is greater than 4000 and is laminar if the Reynolds number is less than 2100.
4. *Plan the solution*: Insert the values into the equation for the Reynolds number and determine whether the flow is laminar or turbulent.
5. *Implement the solution strategy*: The Reynolds number for pipe flow using equation (9.14) is

$$N_{Re} = \frac{\rho \bar{v} d_i}{\mu} = \frac{\left(1032 \frac{kg}{m^3}\right)\left(0.459 \frac{m}{s}\right)(0.02291 \text{ m})}{(1.33 \times 10^{-3} \text{Pa} \cdot \text{s})} = 8,159.5.$$

This high Reynolds number suggests that the flow is turbulent. Convince yourself that all units cancel each other so that Reynolds number is dimensionless (no units).
6. *Check the solution*: In this case, the check should be done by making sure all units are correct and that no math mistakes were made. This answer seems fine.
7. *Generalize the solution strategy*: Calculating the Reynolds number is quite straightforward for Newtonian fluids. For a given fluid (density, viscosity) and pipe, the Reynolds number increases directly with increasing flow velocity.

Practice problem 9.6

It is helpful to understand the effect on the Reynolds number when you alter a given situation, such as changing fluids or pipes.

1. How does the Reynolds number change if the pipe size is increased to 2-in. nominal-size stainless steel sanitary pipe ($d_i = 0.04749$ m) with otherwise the same conditions as in worked problem 9.5? Is the flow still turbulent?
2. How does the viscosity affect the Reynolds number, for example, when a more viscous corn oil ($\rho = 922$ kg/m^3, $\mu = 0.0565$ Pa·s) is flowing in the pipe instead of milk at the velocity from worked problem 9.5? Now is the flow still turbulent?

9.2.3 Flow of a Newtonian fluid in a pipe

The driving force for flow in a pipe is the pressure. That pressure is generally provided either by gravity, in the form of a tank filled with fluid that

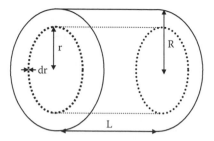

Figure 9.6 Radial slice in a pipe

is raised above the level of the pipe, or by a pump. Newton's second law of motion, which is a form of the conservation of momentum, indicates that the sum of the forces acting on any given fluid element will sum to zero ($\Sigma F_i = 0$). When looking at the steady-state, laminar flow of a radial slice or shell of fluid in a horizontal pipe with a thickness, dr, at radius, r, as shown in Figure 9.6, the forces acting on the shell are the pressure force in, the pressure force out, and the shear force on the circumference. The effect of gravity is considered negligible in a horizontal pipe.

When there is a sufficient distance from the inlet such that the effects on the flow from the fluid entering the pipe are no longer felt and the flow is considered fully developed, the force balance gives

$$P_A \pi r^2 - P_B \pi r^2 - 2\pi r L \sigma = 0. \tag{9.15}$$

Simplifying, the expression becomes

$$\frac{\Delta P}{L} = \frac{2\sigma}{r}. \tag{9.16}$$

For a Newtonian fluid, the shear stress, σ, is defined in equation (9.1). Because the velocity decreases with increasing r, the shear rate within the radial slice is defined as described in section 9.1 as

$$\gamma = -\frac{dv}{dr}. \tag{9.17}$$

Therefore, after substituting and rearranging,

$$\frac{dv}{dr} = -\frac{\Delta P}{2\mu L} r. \tag{9.18}$$

The velocity is assumed to be zero at the pipe wall, which gives a no-slip boundary condition where, at $r = R$, $v = 0$. Therefore, the variables can be separated as described in section 4.3, and this expression can be integrated using

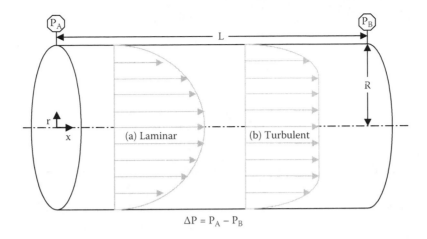

$$\Delta P = P_A - P_B$$

Figure 9.7 Fully developed (a) laminar and (b) turbulent flow profiles in a pipe

integration rules 1 and 2 from section 4.2.2 to get

$$\int_0^v dv = -\frac{P}{2}\frac{}{L}\int_R^r rdr$$

$$v(r) = \frac{\Delta P}{4\mu L}(R^2 - r^2) = \frac{\Delta P R^2}{4\mu L}\left(1 - \frac{r^2}{R^2}\right) \text{ (laminar flow)}. \qquad (9.19)$$

Equation (9.19) describes a parabolic velocity profile for laminar flow such as curve (a) in Figure 9.7. The maximum velocity occurs at the center of the pipe, when $r = 0$.

The volumetric flow rate is found by multiplying the velocity of the fluid in the radial slice shown in Figure 9.6 by the differential cross-sectional area of the slice. Then, integrating across the entire pipe cross-section using integration rules 1, 2, and 3 from section 4.2.2 gives

$$V = \int_0^R v(r)\cdot 2\pi rdr = \frac{\Delta P \pi R^2}{2\mu L}\int_0^R\left(1 - \frac{r^2}{R^2}\right)\cdot rdr$$

$$V = \frac{\Delta P \pi R^4}{8\mu L} \text{ (laminar flow)}. \qquad (9.20)$$

This equation is the one used to find the viscosity of Newtonian fluids in the capillary and pipe viscometers described in section 9.1.2, after solving for the viscosity to get

$$\mu = \frac{\Delta P \pi R^4}{8VL}. \qquad (9.21)$$

The average velocity can be found by dividing the volumetric flow rate by the cross-sectional area to get

$$\bar{v} = \frac{V}{\pi R^2} = \frac{\Delta P R^2}{8\mu L} \text{ (laminar flow)}. \tag{9.22}$$

Equation (9.22) is known as the Poiseuille–Hagen equation.

A typical turbulent flow profile is shown as curve (b) in Figure 9.7. In the case of turbulent flow, the mathematical derivation of the profile is much more complex than that of laminar flow. Therefore, an empirical expression known as the Blasius equation,

$$v(r) = \frac{\Delta P R^2}{4\mu L}\left(1 - \frac{r}{R}\right)^{1/7} \text{ (turbulent flow)} \tag{9.23}$$

is frequently used to represent the velocity profile in turbulent flow. If this expression is used in place of equation (9.19) in equation (9.20), after integration

$$\dot{V} = 0.82\frac{\Delta P \pi R^4}{4\mu L} \text{(turbulent flow)} \tag{9.24}$$

and then

$$\bar{v} = \frac{V}{\pi R^2} = 0.82\frac{\Delta P R^2}{4\mu L} \text{ (turbulent flow)}. \tag{9.25}$$

This shows that the average turbulent flow velocity is 82% of the maximum velocity at the center of the pipe, while the average velocity in laminar flow is just 50% of the maximum velocity in the pipe. This is the result that one would expect based on the shapes of the profiles in Figure 9.7. The average velocity of the flattened turbulent flow profile should be closer to the maximum velocity in the center of the pipe than the average velocity of the parabolic laminar flow profile.

Worked problem 9.7

A capillary tube (4-cm inside diameter, 20 cm long) is used to measure the viscosity of a Newtonian fluid ($\rho = 998$ kg/m^3). When the mass flow rate is 1 kg/s, the pressure drop in the tube is measured at 2.5 kPa. Calculate the viscosity.

Solution:

1. *Motivation*: For Newtonian fluids flowing in a laminar regime, viscosity may be calculated from the pressure drop in a tube with the other

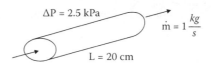

Figure 9.8 Determination of viscosity

fluid properties known. This technique is a fundamental method that may be used in processing conditions to calculate viscosity.

2. *Define the problem*: The fluid properties and flow conditions (pipe size and mass flow rate) are given. We are asked to determine the viscosity, given that the food is Newtonian (Figure 9.8).

3. *Think about the problem*: The mass flow rate may be used to determine the linear fluid velocity. The Reynolds number should be calculated to ensure that the system is laminar. Since viscosity is unknown, this will be used as a check. The Poiseuille–Hagen equation may be used to determine the viscosity.

4. *Plan the solution*: The average fluid velocity will be calculated. This information will be used in the Poiseuille–Hagen equation to calculate viscosity. The viscosity will then be used to calculate the Reynolds number to verify that this is an acceptable method.

5. *Implement the solution strategy*: The mass flow rate in the capillary may be related to the fluid velocity by the following:

$$\dot{m} = \rho\dot{V} = \rho A\bar{v} = \rho\pi R^2\bar{v}. \tag{9.26}$$

Then,

$$\bar{v} = \frac{\dot{m}}{\rho\pi R^2} = \frac{1\,\dfrac{kg}{s}}{\left(998\,\dfrac{kg}{m^3}\right)\pi\left(0.02\,m\right)^2} = 0.797\,\frac{m}{s}.$$

The Poiseuille–Hagen equation (equation [9.22]) is then written as

$$\bar{v} = \frac{\Delta P R^2}{8L\mu} = \frac{\dot{m}}{\rho\pi R^2}. \tag{9.27}$$

Solving for the viscosity and inserting the known values gives

$$\mu = \frac{\Delta P R^2}{8L\mu} = \frac{(2500\,Pa)(0.02\,m)^2}{8(0.2\,m)(0.797\,\tfrac{m}{s})} = 0.784\,Pa\cdot s.$$

The Reynolds number should be calculated using equation (9.14) to ensure that we have laminar flow.

$$N_{\text{Re}} = \frac{\rho \bar{v} d}{\mu} = \frac{\left(998 \frac{\text{kg}}{\text{m}^3}\right)\left(0.797 \frac{\text{m}}{\text{s}}\right)(0.04 \text{ m})}{(0.784 \text{ Pa} \cdot \text{s})} = 40.6$$

This is laminar flow.

6. *Check the solution:* This problem does not have a clear method for checking the solution. We are confident that the Poiseuille–Hagen equation is the right approach.
7. *Generalize the solution strategy:* How might this problem be posed in different ways? One might be asked to solve for any of the variables in the Poiseuille–Hagen equation given the necessary information.

Practice problem 9.8

What is the pressure drop, ΔP, for milk ($\rho = 1032$ kg/m³, $\mu = 0.00133$ Pa $\cdot$ s) that flows 5 m in a pipe ($d_i = 0.02291$ m) with an average velocity of (a) 0.1 m/s or (b) 1 m/s?

9.2.4 Effects of friction on fluid flow

When there is a difference in velocity between two bodies, such as between the stationary pipe and the moving fluid during flow of a fluid food through a pipe between processing operations, some of the energy of motion is lost due to friction. The lost energy is converted to heat that dissipates into the environment. It is accounted for through the use of a friction factor, which is defined as the ratio of the shear stress at the wall, σ_w, to the kinetic energy of the fluid per unit volume:

$$f = \frac{\text{shear stress @ the wall}}{\text{kinetic energy/unit volume}} = \frac{\sigma_w}{\rho \bar{v}^2 / 2}. \tag{9.28}$$

Using equation (9.16) with $r = d_i/2$ gives

$$f = \frac{\Delta P d_i}{2 L \rho \bar{v}^2}. \tag{9.29}$$

In laminar flow, the Poiseuille–Hagen equation (equation [9.22]) can be used to determine the pressure drop and obtain the following expression for the Fanning friction factor:

$$f = \frac{16}{N_{\text{Re}}}. \tag{9.30}$$

Note that there is another form of the friction factor, known as the Darcy friction factor, which is four times the value of the Fanning friction factor. When determining friction-factor values from reference sources, make sure you know which one is being used.

In the cases of transitional and turbulent flow, the mathematical derivation of the friction factor is much less straightforward. In addition, the friction factor is also a function of the surface roughness of the pipe. For most food applications, pipes are made out of materials that can be approximated as a smooth pipe. An empirical expression for the Fanning friction factor in a smooth pipe in turbulent flow is the Blasius equation,

$$f = \frac{0.079}{N_{Re}^{0.25}}. \tag{9.31}$$

The Fanning friction factor can also be determined using the Moody chart, which can be found, for example, in *Perry's Chemical Engineer's Handbook*.

Worked problem 9.9

Calculate the total length of 1-in. nominal, sanitary pipe ($d_i = 0.02291$ m) that would give a ΔP of 70 Pa due to fluid friction for a Newtonian-fluid food flowing at a rate of 0.05 kg/s ($\bar{v} = \dot{m} / \rho A = 0.1213$ m/s), with a viscosity of 2 cP (0.002 Pa·s) and $\rho = 1000$ kg/m³.

Solution:

1. *Motivation*: As fluid foods flow in pipes, the pressure gradient is the driving force for flow. Depending on the fluid, system, and pipe material, there is a certain amount of energy lost in the fluid as it flows through the pipes and fittings. A pump must be used to overcome this fluid friction and supply sufficient energy to keep the fluid moving at the desired rate.
2. *Define the problem*: We are given the fluid and flow properties of the system and are asked to calculate the equivalent length of pipe required to produce the given pressure drop. The equivalent length represents any pipe length, fittings, and process equipment involved in the system (fig. 9.9).

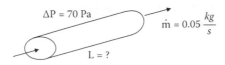

$\Delta P = 70$ Pa $\dot{m} = 0.05 \, \dfrac{kg}{s}$ $L = ?$

Figure 9.9 Equivalent-length calculation

3. *Think about the problem*: The equation for pressure drop with equivalent length of pipe contains the friction factor, f. This is a function of the Reynolds number; therefore, the Reynolds number should be calculated before proceeding with the solution.

4. *Plan the solution*: The Reynolds number will be calculated first. After determining whether the flow is laminar or turbulent, the friction factor will be calculated using the appropriate equation or figure. This will be the last piece needed to solve for the equivalent length through the pressure-drop equation.

5. *Implement the solution strategy*: The Reynolds number should be calculated first using equation (9.14), where the characteristic dimension in the inside diameter of the pipe is

$$N_{Re} = \frac{\rho v d}{\mu} = \frac{\left(1000 \frac{kg}{m^3}\right)\left(0.1213 \frac{m}{s}\right)(0.02291 \text{ m})}{(0.002 \text{ Pa} \cdot \text{s})} = 1389.4.$$

The flow is laminar. Solving for f by substituting the Reynolds number into equation (9.30) gives

$$f = \frac{16}{N_{Re}} = \frac{16}{1389.4} = 0.01152.$$

Now, there is enough information to solve for the equivalent length of pipe. The equivalent length is the length of a straight pipe system that would give the same pressure drop as that in the unknown system. It is found by solving equation (9.29) for L to get

$$L = \frac{\Delta P d_i}{2 \rho \bar{v}^2 f}. \tag{9.32}$$

Substituting the known values into equation (9.32) and calculating the answer gives

$$L = \frac{(70 \text{ Pa})(0.02291 \text{ m})}{2\left(1000 \frac{kg}{m^3}\right)\left(0.12 \frac{m}{s}\right)^2 (0.01152)} = 4.73 \text{ m}.$$

6. *Check the solution*: There is no definite method to check this solution. One should carefully go over calculations to check their accuracy.

7. *Generalize the solution strategy*: We approached this problem by taking the desired parameter and then working backward to determine what pieces of information were needed. We found that two intermediate steps were needed before the solution could be found. This strategy is common to fluid-flow problems, due to the fact that many parameters build on one another.

What is the effect of the various parameters on this calculation? Higher viscosity could result in a lower Reynolds number, higher friction factor, and a shorter equivalent length of pipe needed to give this pressure drop at this flow rate. This makes sense, since more viscous fluids are more difficult to pump.

What are the effects of pipe diameter and flow velocity on this calculation? The following practice problem will help you explore this question.

Practice problem 9.10

What would be the length (a) for a 1.5-in. nominal, sanitary pipe ($d_i = 0.03561$ m) and (b) a flow rate of 0.5 kg/s when all other conditions are the same as in worked problem 9.9?

9.2.5 Mechanical energy balance equation in fluid flow

Ultimately, the flow of fluid foods depends on a source of energy. As we saw in chapter 8, conservation of energy requires that, in steady state, the energy flowing into a system must equal the energy that leaves the system. The most important energy inputs and outputs in fluid flow are kinetic energy ($\bar{v}^2/2\alpha$), potential energy (gz), and pressure (P/ρ) in units of energy per unit mass, where

$\alpha = 1$ for turbulent flow and 0.5 for laminar flow
g = the gravitational constant
z = the height above the reference height.

For steady-state flow of a fluid with no viscosity and a constant density in a pipe with a single inlet at point 1 and an outlet at point 2 where no heat transfer takes place and no work is done, an energy balance gives:

$$\frac{P_1}{\rho} + \frac{\bar{v}_1^2}{2} + gz_1 = \frac{P_2}{\rho} + \frac{\bar{v}_2^2}{2} + gz_2 = \text{constant} \tag{9.33}$$

which is known as the Bernoulli equation. Although real fluids have a viscosity and therefore will have frictional losses, this equation still provides reasonable estimates of unknown variables for fluids with very low viscosities such as water, clear fruit juices, or wine.

When the frictional losses due to the effect of the viscosity are significant, they are represented in the energy balance as a specific frictional energy loss,

$$\frac{P_1}{\mu} + \frac{\bar{v}_1^2}{2} + gz_1 = \frac{P_2}{\mu} + \frac{\bar{v}_2^2}{2} + gz_2 + E_f. \tag{9.34}$$

As discussed in section 2.2.4, there are frictional losses due to flow in the straight pipe sections. These are called the major frictional losses. The major frictional losses for each pipe size in the system are calculated using equation (9.28) to get

$$E_{f,major} = \frac{\Delta P}{\rho} = 2f\frac{\bar{v}^2 L}{D}.$$ (9.35)

When the fluid is flowing through a piping system that contains fittings such as elbows, tees, and valves, or pieces of equipment such as a heat exchanger, there is a larger pressure drop due to friction than can be accounted for by the straight pipe flow. This is due to disruptions in the flow patterns induced in the fluid as it changes direction or speed, or as it passes through the fitting or piece of equipment. The additional frictional losses due to flow effects that occur as food materials flow through fittings or pieces of equipment are called "minor frictional losses." The minor frictional losses for various fittings or pieces of equipment can be accounted for by directly using the experimentally determined pressure drop due to friction across the fitting or piece of equipment. Alternatively, they can be estimated using the expression

$$E_{f,minor} = \frac{\Delta P}{\rho} = C_f\frac{\bar{v}^2}{2}$$ (9.36)

where the frictional loss coefficient, C_f, is also known as the number of velocity heads and is tabulated for a wide variety of fittings in references such as *Perry's Chemical Engineers Handbook* or your textbook. Once the individual losses in each piece of equipment, fitting, and length of pipe are calculated, the total frictional energy loss is the sum of the major and minor frictional losses:

$$E_f = \sum\frac{\Delta P}{\rho} = \sum E_{f,major} + \sum E_{f,minor}.$$ (9.37)

To achieve the desired flow rate, mechanical energy is generally added to a pipe system by a pump or blower. This specific work input (E_p) appears in the mechanical energy balance equation as

$$\frac{P_1}{\rho} + \frac{\bar{v}_1^2}{2} + gz_1 + E_p = \frac{P_2}{\rho} + \frac{\bar{v}_2^2}{2} + gz_2 + E_f.$$ (9.38)

This equation can then be used to determine the energy that needs to be added by a pump to produce a specified flow rate, as demonstrated in the following two problems.

Worked problem 9.11

Milk ($\rho = 1032$ kg/m³, $\mu = 0.00133$ Pa·s) is pumped at an average velocity of 0.25 m/s from one large holding tank at P_{atm} through a heat exchanger to another large holding tank, also at P_{atm}, located 15 m above the level of the first tank. If the flow system provides a total frictional loss (E_f) of 50 m²/s², calculate the pump energy (E_p) required.

Solution:

1. *Motivation*: Calculation of the pump energy required for a flow system is a common goal for any fluid system design problem. The nature of the flow system determines the size of the pump required. The mechanical energy balance, as demonstrated in this problem, is used to calculate pump size.

2. *Define the problem*: The drawing in Figure 9.10 depicts the system. We are given the entrance and exit pressures and height differences, as well as the flow-resistance value. We are asked to calculate the pump energy for this system. The mechanical energy balance equation (9.38) will be necessary to solve the problem.

3. *Think about the problem*: The reference points (points 1 and 2) in equation (9.37) can be located anywhere in the system. The key is to put them in the places that will simplify the problem the most. At the surfaces of both tanks, the pressures are atmospheric and the fluid velocities may be assumed negligible if the surfaces are large. This would eliminate four terms from the mechanical energy equation (velocity and pressure), so we will proceed from there.

4. *Plan the solution*: After choosing the reference points, the given information in the problem may be substituted into the mechanical energy balance equation to solve for the pump energy.

5. *Implement the solution strategy*: The energy requirements of a pump can be calculated by rearranging equation (9.38) to get the expression

$$E_p = \frac{P_2 - P_1}{\rho} + \frac{\bar{v}_2^2 - \bar{v}_1^2}{2\alpha} + g(z_2 - z_1) + E_f. \tag{9.39}$$

From the earlier discussion, the pressure terms are equal and cancel each other, and velocity terms may be set equal to zero. Rearranging gives

$$E_p = g\Delta z + E_f.$$

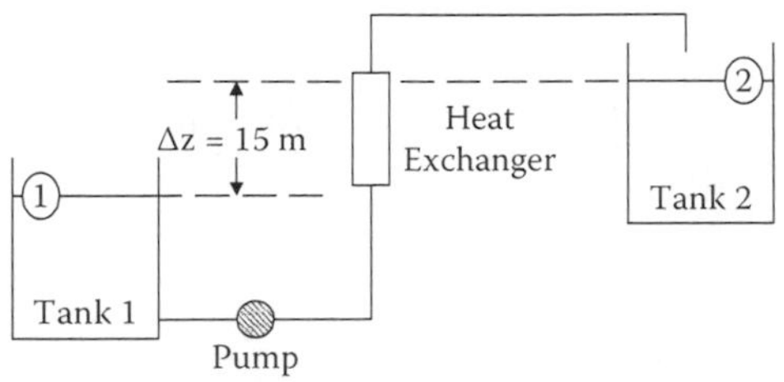

Figure 9.10 Schematic of flow system in problem 9.11

Substituting the values gives

$$E_p = \left(9.8\frac{m}{s^2}\right)(15\ m) + 50\frac{m^2}{s^2} = 197\frac{m^2}{s^2} = 197\frac{J}{kg}.$$

6. *Check the solution*: This answer seems reasonable considering the flow-resistance term.
7. *Generalize the solution strategy*: By choosing appropriate reference points, the mechanical energy balance was simplified greatly. What would happen if we chose a different reference point for the calculation? How might you calculate the pressure at the pump inlet?

Practice problem 9.12

A soft drink with a viscosity of 1.5 cP is being pumped from a holding tank at 1 atm pressure to a pressurized tank at 2 atm of pressure, as shown in Figure 9.11. The tanks are 7 m apart, and the pump is located in between the tanks. The mass flow rate is 5 kg/s, and the pipe has an internal diameter of 0.02664 m. The height difference between the fluid levels in the two tanks is 5 m. The density of the fluid is 1100 kg/m³. Find the work of the pump. Assume frictional losses to be 155 m²/s². (The frictional losses can be calculated for extra practice, using $C_f = 0.5$ to estimate the minor frictional loss as the fluid enters the pipe from tank 1.)

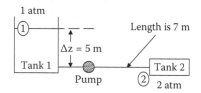

Figure 9.11 Schematic of flow system in problem 9.12

9.3   Non-Newtonian fluid flow

Many fluid foods are non-Newtonian with strongly shear-thinning characteristics. Shear-thinning behavior is generally desirable in pipe systems, since it can significantly reduce the energy required to move fluid foods. In most cases, this behavior can be effectively modeled using the power-law model (equation [9.3]) discussed in section 9.1.1. If this model is used in place of the Newtonian model in equation (9.16), the expression for the velocity profile after integration is

$$v(r) = \left(\frac{\Delta P}{2Lk}\right)^{1/n}\left(\frac{n}{n+1}\right)\left(R^{\left(\frac{n+1}{n}\right)} - r^{\left(\frac{n+1}{n}\right)}\right). \tag{9.40}$$

This expression can then be used to find the volumetric flow rate and average velocity for non-Newtonian power-law fluids by following the same procedure as was done in section 9.2.3 to get

$$\dot{V} = \pi \left(\frac{\Delta \overline{P}}{2Lk} \right)^{1/n} \left(\frac{n}{3n+1} \right) R^{\left(\frac{3n+1}{n} \right)} \text{ (laminar flow)} \qquad (9.41)$$

and

$$\overline{v} = \left(\frac{\Delta P}{2Lk} \right)^{1/n} \left(\frac{n}{3n+1} \right) R^{\left(\frac{n+1}{n} \right)} \text{(laminar flow)}. \qquad (9.42)$$

The pressure drop that will occur over a given length of pipe can then be determined by rearranging equation (9.42) to get

$$\frac{\Delta P}{L} = \frac{4k\overline{v}^n}{D^{(n+1)}} \left(\frac{6n+2}{n} \right)^n. \qquad (9.43)$$

To determine the flow regime, an expression for the Reynolds number that accounts for non-Newtonian power-law effects is needed. If equation (9.42) is used to substitute for the pressure drop per unit length ($\Delta P/L$) in the definition of the Fanning friction factor (equation [9.29]), the resulting expression is the definition of the Fanning friction factor for a non-Newtonian power-law fluid in laminar flow,

$$f = \frac{16}{N_{\text{GRe}}} \qquad (9.44)$$

where

$$N_{\text{GRe}} = \frac{8\rho \overline{v}^{(2-n)} D^n}{k} \left(\frac{n}{6n+2} \right)^n. \qquad (9.45)$$

This generalized Reynolds number (N_{GRe}) is used to define the flow regime for non-Newtonian fluids. The following practice problems demonstrate the determination of the generalized Reynolds number and the use of these non-Newtonian power-law expressions in situations similar to those already encountered in section 9.2 for Newtonian fluids.

Practice problem 9.13

Calculate the generalized Reynolds number, N_{GRe}, for tomato catsup ($\rho = 1130$ kg/m³) flowing at 0.765 m/s in a pipe with $d_i = 0.02291$ m. Catsup is a power-law fluid with a flow behavior index of $n = 0.45$ and a consistency coefficient of $k = 12.5$ Pa·s^n. Calculate the pressure drop per unit length of pipe.

Practice problem 9.14

A non-Newtonian fluid gives a pressure drop of 32 kPa as it flows at 0.5 m/s through a test section of pipe (0.02291 m inside diameter, 2 m long). Calculate the effective viscosity, which may be defined as the viscosity of a non-Newtonian fluid that produces a pressure drop predicted by the Poiseuille–Hagen equation for laminar Newtonian flow.

References

Perry, R. H., and Green, D. 1997. *Perry's Chemical Engineers' Handbook*, 7th ed. New York: McGraw-Hill Professional.

Rao, M. A. 2007. *Rheology of Fluid and Semifluid Foods: Principles and Applications*, 2nd ed. New York: Springer.

Steffe, J. F. 1996. *Rheological Methods in Food Process Engineering*. East Lansing, MI: Freeman Press. Also available online at http://www.freemanpress.com/rheology_book_download.html.

chapter ten

Heat transfer

Heat transfer is arguably the most important transport process in the food industry, since processing of most foods requires either an increase or a decrease in temperature. Thermal processing, or heating to destroy undesirable microorganisms, is one of the main techniques a food technologist utilizes to ensure safety of the food supply. Sterilization and pasteurization require that a food be heated to a certain temperature and held there for sufficient time to destroy microorganisms before being cooled again. Both the heating and cooling steps require an understanding of heat-transfer principles. There are numerous other examples of heat transfer in the food industry, from baking a food in an oven, to solidification of chocolate in a cooling tunnel. Nearly every food processing system utilizes heat transfer in some form or another. Thus, knowledge of the principles of heat transfer is critical for the food technologist. Furthermore, using the principles of heat transfer together with enthalpy balances provides a powerful tool for the food technologist to solve complex food-processing problems.

Heat transfer occurs in one of three modes: conduction, convection, or radiation. Sometimes all three modes of heat transfer may occur at the same time. Conduction heat transfer involves transfer of thermal energy from one molecule to another. A metal rod placed in proximity to a heat source will begin initially to heat up on the end near the heat source. However, after some time, the other end will also be hot, as the molecules of the metal transfer the heat along the length of the rod. Materials that are good heat conductors have high thermal conductivity, k; whereas insulators, materials with low k, do not transfer heat very well by conduction.

Convection heat transfer occurs when a fluid carries thermal energy from one place to another. Heat transfer by fluid motion is much more rapid than conduction, since fluid flow can be quite rapid. Especially in turbulent flow, heat is transferred from a hot source to colder environments quite readily. Convection can either be forced, where the fluid is pumped or circulated to promote heat transfer, or natural, where density variations with temperature result in fluid circulation currents. When fluid is pumped through a plate heat exchanger in pasteurization, forced convection occurs, and heat transfer is quite rapid. Fluid heating in an unstirred pot or kettle will undergo natural convection, where the warmer fluid near the heat source is less dense than the colder fluid farther away from the heat source. Fluid will flow from hot to cold due to the density difference, and this natural convection distributes the heat throughout the container. Circulation in water being heated on a stove is caused by natural convection.

The third mechanism of heat transfer is radiation, which occurs when a hot object radiates heat in the form of electromagnetic radiation. The electromagnetic radiation emitted by a hot object is partially absorbed by a neighboring object, causing that body to heat up. Browning of toast in a toaster is accomplished in part through radiation heat transfer.

Heat transfer can occur under steady-state or unsteady-state conditions. In steady-state heat transfer, the temperature at any point in space is constant. If you were to measure the temperature at any given point in a system, the temperature would be constant in steady-state heat transfer. In unsteady-state heat transfer, temperatures at any point change over time until thermal equilibrium is attained. For example, the temperature of a roast in the oven continues to increase over time until the entire roast reaches the temperature of the oven (thermal equilibrium).

10.1   Steady state

10.1.1   Conduction heat transfer

Heat transfer by conduction takes place through molecular transfer, where the energy is passed from the hot side to the cold side. Fourier's law of heat-transfer conduction is used to describe the rate of heat transfer, $\dot{q}$ (W or J/s), which can be written, in the case of one-dimensional heat transfer, as

$$\dot{q} = -kA\frac{dT}{dx} = -kA\frac{\Delta T}{\Delta x} \tag{10.1}$$

where
$\quad\quad k$ = thermal conductivity (W/m · K)
$\quad\quad A$ = surface area (m^2)
$\quad\quad \Delta T$ = temperature difference from hot to cold sides (°C)
$\quad\quad \Delta x$ = thickness of the material (m).

Materials with a high thermal conductivity and a large surface area transport heat quite well by conduction, especially if they are thin and there is a large temperature difference. In cases where good heat exchange is desired, the goal is to make $\dot{q}$ as large as possible; at other times, low heat transfer (good insulation) is desired. The minus sign in equation (10.1) is simply there to indicate the direction of heat transfer from hot to cold. By mathematical convention, the increase in T is set to coincide with an increase in x, so that heat transfer is in the negative x direction; hence the minus sign.

Problems 10.1 and 10.2 demonstrate the use of Fourier's law of heat transfer in the simplest situations, where temperature difference across a known material is calculated.

Worked problem 10.1

What is the rate of heat transfer per unit area (heat flux, Btu/h · ft² or W/m²) through a 1-cm-thick stainless steel (SS) plate with inside and outside temperatures of 110°C and 75°C, respectively?
Use $k_{SS} = 17$ W/m · °C.

Solution:

1. *Motivation*: This is a standard conduction problem that provides the starting point for beginning the study of heat transfer. Simple conduction heat transfer can be used to describe, for example, heat losses through a wall.
2. *Define the problem*: We are given the thermal conductivity, temperatures at the surfaces, and the thickness of the plate; we are asked to find the rate of heat transfer through the plate. Figure 10.1 will help visualize the problem.
3. *Think about the problem*: Since the temperatures at both sides of the wall are given, we know that the energy loss from the high-temperature side to the low-temperature side must be due to molecular dissipation of energy. This is conduction heat transfer, described by Fourier's law.
4. *Plan the solution*: The conduction equation will be rearranged to give an expression for rate of heat transfer per unit area and solved.
5. *Implement the solution strategy*: The heat-conduction equation is

$$q = -kA \frac{dT}{dx}. \tag{10.1}$$

Multiplying both sides by dx and dividing by A gives

$$\frac{q}{A} dx = -k dT.$$

Since the area is the same on both sides of the wall, A is constant and can be removed from the integral. Integrating from $x = 0$ to 0.01 m

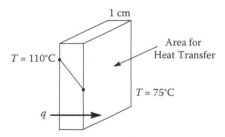

Figure 10.1 Conduction through a wall

(one side of the wall to the other) and $T = 110°C$ to $75°C$, we obtain

$$\frac{\dot{q}}{A} \int_0^{0.01} dx = -k \int_{110}^{75} dT$$

$$\frac{q}{A}(0.01 - 0\text{ m}) = k(110 - 75\ \text{C}).$$

Substituting for k,

$$\frac{q}{A} = \left(17\frac{\text{W}}{\text{mK}}\right)\frac{(35\text{ K})}{(0.01\text{ m})} = 59,500\frac{\text{W}}{\text{m}^2}.$$

6. *Check the solution*: No independent check of this value can be made. However, the units are correct, and the math was done correctly.
7. *Generalize the solution strategy*: This problem required the integration and rearrangement of the conduction heat transfer equation. The same technique would have been necessary for a problem with cylindrical or spherical system geometry; however, in the case of cylindrical or spherical geometry, A would be a function of the dimension of heat transfer. More complicated equations result.

Practice problem 10.2

What is the temperature at the center of the plate in worked problem 10.1?

In most problems of interest in the food industry, complex situations arise where heat transfer occurs across multiple layers of building material. Walls are most often made with multiple layers and even perhaps have multiple sections made of different materials. For example, a wall to a freezer will have inside and outside layers (of wood or metal) for structural integrity, plus at least one layer of insulation between. Furthermore, walls may contain windows where different rates of heat transfer apply. Thus, a distinction must be made when using equation (10.1) as to whether conduction heat transfer occurs in series through multiple resistances, in parallel through multiple sections, or in some combined fashion of series and parallel.

To help solve heat-transfer problems, whether series or parallel, equation (10.1) is often put in a form where heat transfer is induced by a driving force (ΔT) across a resistance ($\Delta x/kA$).

$$\dot{q} = \frac{\text{driving force}}{\text{resistance}} = \frac{\Delta T}{\left(\Delta x/kA\right)} \tag{10.2}$$

This form of the equation is useful for treating complex situations where heat transfer occurs in either series or parallel.

Conduction heat transfer in series means that heat is conducted sequentially through each layer, from the hot side to the cold side. When heat transfers through each section of a multilayer wall in series, the resistances of each element of the wall are summed to give the total resistance,

$$q = \frac{(\Delta T)_t}{\Sigma R_i} = \frac{(\Delta T)_t}{\left(\frac{\Delta x}{kA}\right)_1 + \left(\frac{\Delta x}{kA}\right)_2 + \quad + \left(\frac{\Delta x}{kA}\right)_n}. \tag{10.3}$$

Here, ΔT_t is the total temperature drop across the entire wall, and R_i is the resistance across any of the individual sections. The individual resistances are found from the thickness and thermal conductivity of each element. In this form, multiple resistances can be accounted for in a very simple manner.

In the case of a plane wall, the area of each element is the same, and A can be removed from the denominator of equation (10.3). However, in many problems in the food industry, heat transfer through a cylindrical pipe must be calculated. In this case, the area at the inside radius is different from the area of the outside radius, and A_i must be specified at the appropriate radius. In fact, the proper equation for heat transfer through cylindrical pipes must be derived from the original form of Fourier's law. For heat transfer through the radial dimension of the pipe, Fourier's law is written as

$$\dot{q} = -kA\frac{dT}{dr} \tag{10.4}$$

where r is the radial dimension of the pipe. In this case, the surface area of the pipe depends on r according to

$$A = 2\pi rL \tag{10.5}$$

where L is the length of the pipe. Substituting equation (10.5) into equation (10.4) gives

$$q = -2\pi kLr\frac{dT}{dr}. \tag{10.6}$$

This simple differential equation can be solved by separation of variables and integration across the appropriate boundary conditions. Separating variables (see section 4.3) gives

$$q\frac{dr}{r} = -2\pi kLdT. \tag{10.7}$$

Integrating from T_1 at $r = 0$ to T_2 at an arbitrary r gives

$$\dot{q} \int_0^r \frac{dr}{r} = -2\pi kL \int_{T_1}^{T_2} dT \tag{10.8}$$

$$q \ln\left(\frac{r_2}{r_1}\right) = -2\pi kL(T_2 - T_1).$$

Rearranging gives

$$q = -2\pi kL \frac{\Delta T}{\ln\left(\frac{r_2}{r_1}\right)}. \tag{10.9}$$

Equation (10.9) is used to calculate the rate of heat transfer across the wall of a pipe. When there are multiple layers of pipe, as for example would occur when a layer of insulation is placed around a metal pipe, then equation (10.9) is rewritten in the form of multiple resistances

$$\dot{q} = -2\pi L \frac{\Delta T_t}{\left\{ \dfrac{\ln\left(\frac{r_2}{r_1}\right)}{k_1} + \dfrac{\ln\left(\frac{r_3}{r_2}\right)}{k_2} + \dfrac{\ln\left(\frac{r_4}{r_3}\right)}{k_3} \right\}} \tag{10.10}$$

where r_1 through r_4 represent the different radii associated with the multiple layers. For example, in this case, there are three layers of material (perhaps one thickness of pipe and two layers of insulation). If additional resistances are needed, they are simply added in the brackets in the denominator.

Problems 10.3 and 10.4 provide examples of multilayered walls with heat-transfer resistances in series.

Worked problem 10.3

A wall is made of 2-in. concrete ($k = 0.54$ Btu/ft·h·°F) with 2 in. of corkboard insulation ($k = 0.025$ Btu/ft·h·°F). If the temperature at the inside surface is 35°F and at the outside surface is 85°F, find the rate of heat transfer per ft^2 of wall.

Solution:

1. *Motivation*: Heat transfer through walls of composite materials is a major concern in the food industry. Heat transfer through a pipe and insulation is an example of a situation where these concepts apply.
2. *Define the problem*: We are given the wall thickness and thermal conductivities, as well as the outside surface temperatures. We are asked to find the rate of heat transfer through the wall (Figure 10.2).

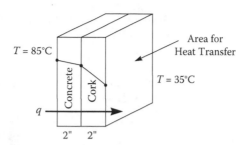

Figure 10.2 Conduction through a composite wall

3. *Think about the problem:* The rate of heat transfer through each section is always the same since this is steady-state heat transfer. The conduction equation for composite walls (equation 10.2) may be applied here.

4. *Plan the solution:* We will use the conduction equation for composite walls with two resistances to heat transfer. One resistance is due to the concrete, and the other is due to the thickness of the cork insulation.

5. *Implement the solution strategy:* The equation for the rate of heat transfer in composite walls with two conduction resistances is

$$\dot{q} = \frac{A\Delta T_{total}}{\left[\left(\frac{\Delta x}{k}\right)_{concrete} + \left(\frac{\Delta x}{k}\right)_{cork}\right]} \tag{10.11}$$

where

ΔT_{total} = the temperature change across both layers (°F)
Δx_i = the thicknesses of the individual layers (m)
k_i = the individual conductivities (Btu/ft·h·°F)
A = the heat transfer surface area (m²).

Now, substitute the necessary information, being careful to make sure units match.

$$\frac{\dot{q}}{A} = \frac{(85°F - 35°F)}{\left[\left(\frac{\frac{2}{12}\,ft}{0.54\,\frac{Btu}{ft·h·°F}}\right) + \left(\frac{\frac{2}{12}\,ft}{0.025\,\frac{Btu}{ft·h·°F}}\right)\right]} = \frac{50°F}{\left[0.309\,\frac{ft^2·h·°F}{Btu} + 6.66\,\frac{ft^2·h·°F}{Btu}\right]}$$

$$\frac{\dot{q}}{A} = 7.2\,\frac{Btu}{ft^2·h}$$

6. *Check the solution:* Again, no independent check of this value can be made.

7. *Generalize the solution strategy:* Any composite wall with series resistances can be solved in this manner. If several more layers were present in the wall, equation (10.11) would be rewritten to include the additional resistance terms.

In this problem, heat transfer occurred through resistances in series, where heat transferred first through one and then the next layer. Suppose there was a window in this wall, where heat transferred through the wall and the window at the same time. How would you approach this problem? We will address this problem in the next section on heat transfer in parallel systems.

Practice problem 10.4

Calculate the rate of heat transfer through a 10-m section of pipe (ID = 0.02291 m; OD = 0.0254 m) made of stainless steel ($k = 17$ W/m · °C) if temperatures at the inside and outside surfaces are 100°C and 25°C, respectively.

In the previous examples, the resistances were in series, meaning that heat was transferred through each of them sequentially. In some cases, heat transfer occurs in parallel. A window in a wall is an excellent example of resistances in parallel. Heat transfer through both the wall and window can be calculated individually according to equation (10.3). However, now the total amount of heat transferred through the combined window/wall is the sum of the heat passing through the wall and the window. These are resistances in parallel, and the total heat transfer is then

$$\dot{q}_t = \dot{q}_{win} + \dot{q}_{wall} \tag{10.12}$$

where q_{win} is the rate of heat transfer through the window, and q_{wall} is the rate of heat transfer through the wall (minus the window). If there are multiple windows (or multiple resistances in parallel), these are summed to give the total rate of heat transfer through the complex wall. Problem 10.5 provides an example of heat transfer through resistances in parallel.

Practice problem 10.5

A wall (90 ft^2) to a refrigerator has an observation window (10 ft^2) installed. Compare the rate of heat transfer of the original wall (no window) to the new wall (with window). The wall is made up of 1/8-in. metal ($k = 28.9$ Btu/ft·h·°F) on the inside, a 1-in. layer of corkboard as insulation ($k = 0.025$ Btu/ft·h·°F), and a 1-in. layer of fiberboard ($k = 0.028$ Btu/ft·h·°F). The window is a single pane of glass ($k = 0.6$ Btu/ft·h·°F). The temperature difference across the wall is 50°F.

10.1.2   Convection heat transfer

In the previous examples of conduction heat transfer, it was assumed that the temperature directly at the surface was known and that convection (related to fluids) was not involved. However, in reality, wall temperatures are rarely known, but rather the bulk fluid temperatures on each side of a wall are known. For example, in problem 10.6, a temperature difference across the wall of 50°F was specified and this was assumed to be the temperature difference exactly at the inside and outside of the wall. In reality, though, it is the air temperature on both sides that is known, and there is convection heat transfer between the bulk air and the surface of the wall. Thus, an additional resistance to heat transfer on each side of the wall is needed.

Heat transfer by convection is calculated from

$$\dot{q} = hA\Delta T \tag{10.13}$$

where h is the convective heat-transfer coefficient. The value of h can be found from equations that correlate flow conditions with the geometry of the system under consideration (to be covered in the next section). In complex systems where both conduction and convection occur, the rate of heat transfer can be calculated by summing all of the resistances. For a wall with a single element and convective coefficients on each side, heat transfer can be calculated from

$$\dot{q} = \frac{\Delta T_t}{\left\{ \frac{1}{h_i A} + \frac{\Delta x}{kA} + \frac{1}{h_o A} \right\}} \tag{10.14}$$

where h_i and h_o are the convective heat-transfer coefficients on the inside and outside of the wall, respectively. If there are more elements to the wall, then additional resistance elements would be summed within the braces of equation (10.14) as needed. For a flat wall, the area terms, A, are all the same and can be removed from the braces.

$$\dot{q} = A \frac{\Delta T_t}{\left\{ \frac{1}{h_i} + \frac{\Delta x}{k} + \frac{1}{h_o} \right\}} \tag{10.15}$$

Equation (10.15) is often simplified by taking the terms within the braces as an overall heat-transfer coefficient, U. The heat-transfer equation for multiple resistances (combined convection and conduction) is written as

$$q = UA\Delta T_t. \tag{10.15a}$$

The term, U, combines all the heat-transfer resistances from conduction and convection for resistances in series.

For cylindrical tubes, the definition of U is slightly different, since the area defined for each heat-transfer element is different. The total rate of heat transfer for a pipe with convective resistances on both sides is

$$\dot{q} = \frac{\Delta T_t}{\left\{ \frac{1}{h_i A_i} + \frac{\ln\left(\frac{r_o}{r_i} \right)}{2\pi kL} + \frac{1}{h_o A_o} \right\}} \tag{10.16}$$

where A_i and A_o are the surface areas for heat transfer at the inside and out-side surfaces (r_i and r_o) of the pipe, respectively. To rewrite this equation in the same form as equation (10.14) requires removal of one of the area terms from the brackets and the definition of the resulting overall heat-transfer coefficient according to that area. Most commonly, the inside area of a pipe ($A_i = 2\pi r_i L$) is chosen, and equation (10.16) can be rewritten as

$$\dot{q} = U_i A_i \Delta T_t \tag{10.17}$$

where

$$\frac{1}{U_i} = \frac{1}{h_i} + \frac{r_i \ln\left(\dfrac{r_o}{r_i}\right)}{2\pi k L} + \frac{r_i}{h_o r_o}. \tag{10.18}$$

Worked problem 10.6

A refrigerator wall is made of 1-in.-thick plywood ($k = 0.067$ Btu/ft·h·°F), a 1-in. layer of insulation ($k = 0.019$ Btu/ft·h·°F), and 1 in. of fiberboard ($k = 0.33$ Btu/ft·h·°F). Temperature inside should be maintained at 35°F, and outside temperature is typically 70°F. Air movement, both in the refrigerator and along the outside wall, is slight, so that convective heat-transfer coefficients are $h_i = 2$ Btu/ft²·h·°F and $h_o = 5$ Btu/ft²·h·°F. Calculate the overall heat transfer coefficient, U, and the rate of heat loss (per ft² of area) into the refrigerator through the walls.

Solution:

1. *Motivation*: The overall heat-transfer coefficient is an integral part of heat transfer. This coefficient combines the effects of convection and conduction and is an important measure of how fast heat is transferred within a system. The rate of heat transfer through the walls in this problem is one of the factors that influences how much refrigeration we need. What other factors influence the amount of refrigeration needed?
2. *Define the problem*: We are given the makeup of the refrigerator wall (Figure 10.3). The thicknesses and thermal conductivities of the materials are provided, as well as the temperature environment and heat-transfer coefficients on the inside and outside of the door. We are asked to find the overall heat-transfer coefficient and the heat loss through the door.
3. *Think about the problem*: This is a composite wall with three conduction and two convection heat-transfer resistances in series. Thus, the rate of heat transfer through each section must be the same in order to have steady-state heat transfer (temperature constant with time). Since both conduction and convection heat transfer occur, the overall heat-transfer equation must include both types of resistances.

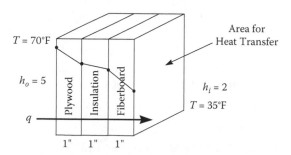

Figure 10.3 Refrigerator door

4. *Plan the solution*: The overall heat-transfer coefficient for heat transfer in a series through a plane can be calculated by summing the individual heat-transfer resistances for each section. Once the overall heat-transfer coefficient is known, we can calculate the rate of heat transfer through the entire wall.

5. *Implement the solution strategy*: The overall heat-transfer coefficient can be calculated by summing the individual resistances for convection (inside and outside) and conduction (layers of plywood, insulation, and fiberboard).

$$\frac{1}{U} = \frac{1}{h_i} + \left(\frac{\Delta x}{k}\right)_{ply} + \left(\frac{\Delta x}{k}\right)_{ins} + \left(\frac{\Delta x}{k}\right)_{fiber} + \frac{1}{h_o} \qquad (10.19)$$

Substituting the given values, one obtains

$$\frac{1}{U} = \frac{1}{2\dfrac{Btu}{ft^2 \cdot h \cdot {}^\circ F}} + \left(\frac{\frac{1}{12}\,ft}{0.067\,\frac{Btu}{ft \cdot h \cdot {}^\circ F}}\right) + \left(\frac{\frac{1}{12}\,ft}{0.019\,\frac{Btu}{ft \cdot h \cdot {}^\circ F}}\right) + \left(\frac{\frac{1}{12}\,ft}{0.33\,\frac{Btu}{ft \cdot h \cdot {}^\circ F}}\right) + \frac{1}{5\,\frac{Btu}{ft^2 \cdot h \cdot {}^\circ F}}$$

$$U = 0.1519\,\frac{Btu}{ft^2 \cdot h \cdot {}^\circ F}.$$

The rate of heat transfer per unit area is given by

$$\frac{q}{A} = U\Delta T = \left(0.1519\,\frac{Btu}{ft^2 \cdot h \cdot {}^\circ F}\right)(70 - 35^\circ F) = 5.3\,\frac{Btu}{ft^2 \cdot h}.$$

6. *Check the solution*: No independent check of this answer can be done. Does it seem too high or too low?

7. *Generalize the solution strategy*: It is sometimes necessary to find the temperatures at the points between resistances. How would one go about calculating the temperatures at each of the wall surfaces in this problem?

Practice problem 10.7

In the problem above, the air within the refrigerator is now moving rapidly (a fan is blowing), so that $h_i = 100$ Btu/ft$^2 \cdot$ h $\cdot$ °F. How much greater is the overall heat-transfer coefficient and the rate of heat transfer?

Worked problem 10.8

A stainless steel pipe ($k = 17$ W/m $\cdot$ K) containing hot condensate at 100°C is insulated by a 1-cm-thick layer of insulation ($k = 0.015$ W/m $\cdot$ K). The pipe has $d_i = 0.03561$ m and $d_o = 0.0381$ m. Still air on the outside of the pipe at 25°C results in $h_o = 10$ W/m$^2 \cdot$ K, and condensate flow within the pipe gives $h_i = 200$ W/m$^2 \cdot$ K. Find the overall heat transfer coefficient, U, and the rate of heat transfer per unit length of pipe.

Solution:

1. *Motivation*: Again, calculating the overall heat-transfer coefficient for a food process may be the most important part of the heat-transfer analysis. This type of analysis allows one to calculate the amount of insulation needed to minimize heat loss from a hot pipe.
2. *Define the problem*: The pipe and insulation properties (diameters, thicknesses, and thermal conductivities) are given, as well as the convective heat-transfer coefficients on the inside and outside of the pipe (Figure 10.4). We are asked to calculate the overall heat-transfer coefficient and the heat loss per unit length of pipe given the temperature difference.
3. *Think about the problem*: This problem is similar to problem 10.6, except that this one is in cylindrical coordinates. The overall heat-transfer coefficient may be calculated by summing the heat-transfer resistances in series using the equation for heat transfer in cylindrical coordinates. In this case, we want to find rates of heat transfer per

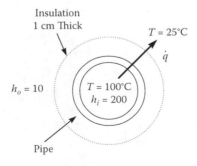

Figure 10.4 Conduction through composite pipe

unit length of pipe $(\dot{q}/L)$. To get $(\dot{q}/L)$, we need to write out the surface area, A, as πdL and divide by L.

4. *Plan the solution*: There are two possibilities for solving this problem: using the overall heat-transfer coefficient based on the inside surface area (U_i) or based on the outside surface area (U_o). For this problem we will choose the inside surface area for our basis. Again, we find U_i by summing the individual resistances: the inside and outside convective resistances and conduction resistances through the pipe and insulation.

5. *Implement the solution strategy*: The overall heat-transfer coefficient based on the inside surface area for cylindrical coordinates and systems with two conduction resistances is

$$\frac{1}{U_i} = \frac{1}{h_i} + \frac{r_i \ln\left(\frac{r_{o,pipe}}{r_i}\right)}{k_{pipe}} + \frac{r_i \ln\left(\frac{r_{o,insulation}}{r_{o,pipe}}\right)}{k_{insulation}} + \frac{r_i}{h_o r_{o,insulation}}. \tag{10.20}$$

Substituting the known values gives

$$\frac{1}{U_i} = \frac{1}{200\frac{W}{m^2K}} + \frac{(0.017805 \text{ m})\ln\left(\frac{0.01905 \text{ m}}{0.017805 \text{ m}}\right)}{17\frac{W}{mK}} + \frac{(0.017805.)\ln\left(\frac{0.02905 \text{ m}}{0.01905 \text{ m}}\right)}{0.015\frac{W}{mK}}$$

$$+ \frac{0.017805 \text{ m}}{\left(10\frac{W}{m^2K}\right)(0.02905 \text{ m})}$$

$$\frac{1}{U_i} = 0.5672\frac{m^2K}{W}$$

$$U_i = 1.763\frac{W}{m^2K}.$$

Now, the rate of heat transfer is

$$q = U_i A_i \Delta T = U_i(\pi d_i L)\Delta T. \tag{10.21}$$

Rearranging by dividing by the pipe length gives

$$\frac{q}{L} = U_i \pi d_i \Delta T = \left(1.763\frac{W}{m^2K}\right)(\pi(0.03561 \text{ m}))((100-25)K) = 14.8\frac{W}{m}.$$

6. *Check the solution*: This solution may be checked by calculating the overall heat-transfer coefficient based on the outside surface area and the rate of heat transfer per unit length with the appropriate equation. In this case,

$$U_o = 1.0806\frac{W}{m^2K}.$$

This should be different than the inside overall heat transfer coefficient, but the rate of heat transfer should be the same.

$$\frac{\dot{q}}{L} = U_o \pi d_o \Delta T = \left(1.0806 \frac{W}{m^2 K}\right)(\pi(0.0581\ m))((100-25)K) = 14.8 \frac{W}{m}$$

The answer is the same as before.

7. *Generalize the solution strategy*: In later calculations, we will need to calculate the temperature of the inside surface of the pipe. How might you calculate that temperature based on heat-transfer principles?

Practice problem 10.9

The pipe in problem 10.8 develops a layer of scale on the inside due to the hot condensate, and this fouling layer can be characterized by a fouling coefficient, h_f, of 100 W/m$^2 \cdot$ K. How much do the overall heat-transfer coefficient and rate of heat transfer decrease?

10.1.3 Convective heat-transfer coefficients

Convective heat-transfer coefficients are estimated from empirical correlation equations. To get these equations, numerous experiments measuring the rate of heat transfer under controlled conditions allow correlation of h with operating parameters like flow rates and geometry. A typical correlation equation is given as a correlation between the Nusselt number (hD/k), the Reynolds number ($\rho v D/\mu$), and the Prandtl number ($\mu c_p/k$).

$$N_{Nu} = a + b(N_{Re})^c (N_{Pr})^e \tag{10.22}$$

where a, b, c, and e are best-fit parameters for the particular model. Such correlation equations have been found for a wide variety of geometries applicable to the food industry. A wide range of heat-transfer correlation equations can be found in *Perry's Chemical Engineer's Handbook*.

One correlation of great importance to the food industry is for flow in pipes or channels. As a fluid flows through a heat exchanger, heat is transferred from the hot side to the cold side according to convective and conductive heat-transfer mechanisms. The rate of heat transfer and the applicable correlation equation for flow in pipes depends on the flow regime. For laminar flow in cylindrical pipes, the correlation equation is

$$N_{Nu} = 1.86 \left(N_{Re} N_{Pr} \frac{D}{L}\right)^{1/3}. \tag{10.23}$$

For turbulent flow in pipes, the following correlation equation is often used

$$N_{Nu} = 0.023(N_{Re})^{0.8}(N_{Pr})^{1/3}. \tag{10.24}$$

To calculate the convective heat-transfer coefficient for a food processing operation, the correct correlation equation must be used. But, since these equations are based on experimental correlations, the values of h found from these correlation equations are typically only accurate within 5–20%. That is, the calculation of h from correlation equations for any given situation only gives an estimation of the actual value. This ambiguity of convective heat-transfer coefficients is one of the main reasons that engineers normally overdesign equipment, especially heat exchangers. Equipment manufacturers generally have correlation equations for their specific equipment that allow them to calculate h-values more accurately; however, manufacturer's equations are generally considered proprietary.

Problems 10.10 through 10.13 provide examples of calculation of h for various conditions.

Worked problem 10.10

Air (at 2°C) blowing across an orange at 2 m/s (Figure 10.5) cools the orange to refrigerator temperature. The convective heat-transfer coefficient, h, for flow across a sphere is given by

$$N_{Nu} = 2 + 0.6(N_{Re})^{1/2}(N_{Pr})^{1/3} \tag{10.25}$$

$$\frac{hd}{k} = 2 + 0.6\left(\frac{\rho \bar{v} d}{\mu}\right)^{1/2}(N_{Pr})^{1/3} \tag{10.26}$$

where d is the diameter of the sphere, and this equation applies for all N_{Re} between 1 and 70,000.

Note that physical properties of the air are determined at the film temperature, midway between product surface temperature and the air temperature. If the orange is initially at 25°C and is 8 cm in diameter, calculate h during the initial stages of cooling. For air, use: $\rho = 1.206$ kg/m³; $k = 0.0244$ W/m·K; $N_{Pr} = 0.71$; and $\mu = 1.7848 (10^{-5})$ Pa·s. These values were found in a table for properties of air, with a temperature of 13.5°C.

Solution:

1. *Motivation:* Convective heat-transfer coefficients are required to solve most heat-transfer problems. However, these values must be estimated

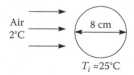

Air
2°C
8 cm
$T_i = 25°C$

Figure 10.5 Cooling of an orange

from correlations such as equation (10.22), based on the flow conditions and physical system of interest. The food scientist should understand how to use the correlation equations to calculate these coefficients.

2. *Define the problem*: Air is blowing across the orange, as shown in Figure 10.5. The Prandtl number is given, as well as the parameters for the Reynolds number. The thermal conductivity and radius of the orange are also given so that we may calculate the convective heat-transfer coefficient.

3. *Think about the problem*: This problem requires the calculation of the Nusselt number before the convective heat-transfer coefficient can be calculated. The Reynolds number is necessary to calculate the Nusselt number.

4. *Plan the solution*: First, the Reynolds number for the air around the orange must be calculated so that the Nusselt number can be calculated. After this, the convective heat-transfer coefficient can be calculated.

5. *Implement the solution strategy*: The Reynolds number for the air flowing around the sphere is

$$N_{\text{Re,sphere}} = \frac{\rho v d}{\mu}. \tag{10.27}$$

Substituting the above values into the equation gives

$$N_{\text{Re,sphere}} = \frac{\left(1.206\,\frac{\text{kg}}{\text{m}^3}\right)\left(2\,\frac{\text{m}}{\text{s}}\right)(0.08\,\text{m})}{(1.7848(10^{-5})\text{Pa}\cdot\text{s})} = 10{,}811.$$

This is within the range of values for equation (10.25). The Nusselt number may be calculated from the correlation by substituting the Reynolds number and Prandtl number,

$$N_{\text{Nu}} = 2 + 0.6(10{,}811)^{\frac{1}{2}}(0.71)^{\frac{1}{3}} = 57.65.$$

Now, the convective heat-transfer coefficient may be calculated by

$$h = \frac{N_{\text{Nu}}k}{d} = \frac{(57.65)\left(0.0244\,\frac{\text{W}}{\text{mK}}\right)}{0.08\,\text{m}} = 17.58\,\frac{\text{W}}{\text{m}^2\text{K}}.$$

Air blowing across the orange at 2 m/s gives a convective heat-transfer coefficient of about 17.6 W/m²·K.

6. *Check the solution*: Does this seem like a reasonable number? You should know that convective coefficients are generally quite low for air cooling, so a value of 17.6 W/m²·K seems reasonable.

7. *Generalize the solution strategy*: Correlations for heat-transfer coefficients are common and necessary in food applications. How does increased air velocity affect the convective heat-transfer coefficient? It

raises the Reynolds number, which raises the Nusselt number, which raises the convective heat transfer coefficient. How does "wind chill" in cold weather relate to convection?

Practice problem 10.11

Calculate the convective heat-transfer coefficient for natural convection of air (25°C) above a flat heated plate (at 100°C). The correlation equation for this case is $h = C(\Delta T)^{0.25}$, where $C = 2.4493$ and ΔT is the temperature difference between the surface of the plate and the air.

Worked problem 10.12

What would h be if the air flow across the orange in problem 10.10 was only 0.2 m/s? Comment.

Solution:

1. *Motivation*: This problem addresses how convection varies with different flow regimes in cooling or heating problems. One of the primary factors that affects convection is the flow rate of the heating or cooling medium. We need to understand this influence.
2. *Define the problem*: This problem is identical to problem 10.10, except that the air velocity is one order of magnitude smaller.
3. *Think about the problem*: This reduced air velocity should lower the Reynolds number, which lowers the Nusselt number. Therefore, we should find a lower convective heat-transfer coefficient.
4. *Plan the solution*: Same as problem 10.10.
5. *Implement the solution strategy*: The new Reynolds number is

$$N_{Re,sphere} = \frac{\left(1.206\,\frac{kg}{m^3}\right)\left(0.2\,\frac{m}{s}\right)(0.08\ m)}{(1.7848(10^{-5})Pa\cdot s)} = 1081.$$

This is still within the range of values, which makes equation (10.15) valid. The Nusselt number may be calculated from the correlation by substituting the Reynolds number and Prandtl number.

$$N_{Nu} = 2 + 0.6(1081)^{1/2}(0.71)^{1/3} = 19.60$$

Now, the convective heat-transfer coefficient can be calculated by

$$h = \frac{N_{Nu}k}{d} = \frac{(19.60)\left(0.0244\,\frac{W}{m\cdot K}\right)}{0.08\ m} = 5.98\,\frac{W}{m^2\cdot K}.$$

This value is roughly one-third of the value when the velocity was 2 m/s.

6. *Check the solution*: A decrease in air velocity by a factor of 10 caused a decrease in h by about a factor of 3. At the lower air velocity, we get a lower convective heat-transfer coefficient. This sounds reasonable.

7. *Generalize the solution strategy*: The lower air velocity lowered the convective heat-transfer coefficient, as expected. What is the relationship between air velocity and h? Calculate h for several different air velocities and plot to see this relationship. Does h approach some limiting value when v is very high or low? Why might this occur? At high velocities, is h a function of the square root of the velocity?

Practice problem 10.13

Now calculate h for water flowing across the orange (previous problems). For water, use $v = 0.2$ m/s; $\rho = 999.7$ kg/m^3; $k = 0.577$ W/m $\cdot$ K; $\mu = 1.296\,(10^{-3})$ Pa $\cdot$ s; and $N_{Pr} = 9.5$. What can you conclude about the difference in water and air as heat-transfer media?

10.1.4   Heat exchangers

A common process operation in the food industry involves heat transfer between two fluids. Pasteurization of milk, for example, is accomplished in a plate heat exchanger, with the raw milk being heated to pasteurization temperature by use of a hot fluid. Heat transfer between the hot and cold fluids is based on the flow properties of the system and the heat-transfer coefficient across the wall of the exchanger. A combination of conduction and convection heat transfer applies in this case. Thus, the rate of heat transfer can be calculated from equation (10.16), for tube geometry, based on knowledge of the overall heat-transfer coefficient and the appropriate temperature difference as the driving force.

In a heat exchanger, however, the temperature-difference driving force changes along the length of the heat exchanger. The exact nature of changing temperature depends on the direction of flow of each of the fluids in the heat exchanger. Perhaps the simplest heat exchanger is a double-tube arrangement, where one tube fits concentrically into a second, larger tube. One fluid flows within the inside tube, and the other fluid flows in the annular space between the two tubes. If both fluids flow in the same direction, the flow pattern is cocurrent, and one fluid heats up as the other fluid cools down. Eventually, the two fluids reach the same temperature, and no further heat transfer occurs. The limitation to cocurrent heat-exchanger operation is that, as the two fluids get closer in temperature, the driving force for heat transfer decreases, and the rate of heat transfer eventually goes to zero. Thus, it is more common for the two fluids to flow in opposite directions, or in countercurrent operation. In this case, the hot fluid gives up its heat to warm the cold fluid, but there is no limitation on the end point of heat transfer. That is, for example, the hot fluid can come out colder than the cold fluid comes out.

Regardless of which direction the two fluids flow, the principles of heat transfer and enthalpy balances still apply. However, since the driving force for heat transfer changes along the length of the heat exchanger, an average temperature difference is used. Rather than using an arithmetic average, however, it is technically more correct to use a logarithmic mean temperature, as defined by

$$(\Delta T_{\text{lm}}) = \frac{\ln(\Delta T_2) - \ln(\Delta T_1)}{\ln\left(\frac{\Delta T_2}{\Delta T_1}\right)} \tag{10.28}$$

where ΔT_{lm} is a logarithmically averaged temperature (log mean temperature difference) along the length of the heat exchanger. The values for ΔT_1 and ΔT_2 are the temperature differences between hot and cold fluids at each end of the heat exchanger.

Because the temperature driving force ($T_{\text{hot}} - T_{\text{cold}}$) changes, the rate of heat transfer also changes along the length of the heat exchanger. The average rate of heat transfer is then obtained by using ΔT_{lm} as the average driving force,

$$q = U_i A_i \Delta T_{\text{lm}}. \tag{10.29}$$

The enthalpy balance on the heat exchanger can be written based on the flow rates of the two fluids. Assuming steady-state operation with no heat losses, the enthalpy balance is

$$\dot{q} = (\dot{m}c_p \Delta T)_1 = (\dot{m}c_p \Delta T)_2 \tag{10.30}$$

where flow rates, m, and specific heats, c_p, are determined for the hot and cold fluids passing through the heat exchanger. The values of ΔT used for the enthalpy balance are the differences between inlet and outlet temperatures for each fluid and should not be confused with ΔT_{lm}, the driving force for heat transfer across the wall of the heat exchanger.

To solve problems associated with heat exchangers, equations (10.29) and (10.30) are combined, since the rate of heat transfer in both equations should be the same to give steady-state operating conditions. These equations can be used, for example, to calculate the required area of a heat exchanger to give the desired cooling or heating effects. Alternatively, these equations can be used to determine whether an existing heat exchanger will accomplish the desired effect. Problems 10.14 and 10.15 demonstrate the use of heat-transfer and enthalpy-balance principles for heat-exchanger problems.

Worked problem 10.14

Milk (c_p = 3817 J/kg·K) flowing at 0.1 kg/s is heated from 4°C to 30°C in a tubular heat exchanger (countercurrent flow) using hot water, which cools

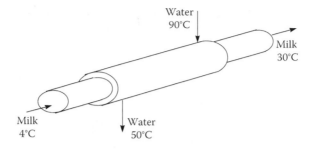

Figure 10.6 Schematic of countercurrent heat exchanger

from 90°C to 50°C. Calculate the required length of heat exchanger for $h_i = 500$ W/m² · K; $h_o = 700$ W/m² · K; $d_o = 2.54$ cm; $d_i = 2.291$ cm; $k_{pipe} = 17.3$ W/m · K.

Solution:

1. *Motivation*: Heat exchangers are the primary equipment that heats or cools a fluid food during processing. Each heat exchanger is designed uniquely to heat or cool a food to a desired temperature. To solve problems on heat exchangers, you must know how to combine enthalpy balances with heat-transfer calculations.

2. *Define the problem*: We are given the temperature increase of the milk, the heat-transfer coefficient, and the specific heat of the milk. We are also given the temperature decrease and heat-transfer coefficient of the hot water. Lastly, the properties of the tubular heat exchanger are also given. We are asked to calculate the length of the heat exchanger. Figure 10.6 is a schematic of the heat exchanger, and Figure 10.7 shows how the temperature changes over the length of the exchanger.

3. *Think about the problem*: Heat is transferred from the water to the milk. The rate of heat transfer to the milk may be calculated from the enthalpy

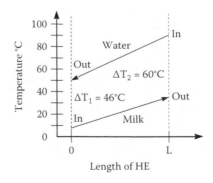

Figure 10.7 Temperature changes for milk and water in countercurrent heat exchanger

balance on the milk. This must be the same rate of heat transfer calculated through the heat-exchanger equation in order to have steady state. The overall heat-transfer coefficient of the exchanger may be calculated from the sum of convective and conductive resistances. Since the driving force for heat transfer ($T_{hot} - T_{cold}$) changes along the length of the heat exchanger, we will need to use an average value (log mean temperature difference). The length may then be calculated from the area of the heat exchanger required to produce these temperatures.

4. *Plan the solution*: We will calculate the rate of heat transfer from the enthalpy balance. The overall heat-transfer coefficient and log mean temperature difference will be calculated. The surface area of heat transfer can then be determined, and thus the length can be calculated.

5. *Implement the solution strategy*: The rate of heat transfer to the milk may be calculated as

$$q = mc_p \Delta T = \left(0.1\frac{\text{kg}}{\text{s}}\right)\left(3817\frac{\text{J}}{\text{kg}\cdot\text{K}}\right)(30-4^\circ\text{C}) = 9924.2 \text{ W}.$$

The overall heat-transfer coefficient based on the inside surface area (U_i) is

$$\frac{1}{U_i} = \frac{1}{h_i} + \frac{r_i\ln\left(\frac{r_{o_i}}{r_i}\right)}{k} + \frac{r_i}{h_o r_o} = \frac{1}{500\frac{\text{W}}{\text{m}^2\cdot\text{K}}} + \frac{(.01146\text{ m})\ln\left(\frac{0.0172\text{ m}}{0.01146\text{ m}}\right)}{17.3\frac{\text{W}}{\text{m}\cdot\text{K}}}$$

$$+ \frac{0.01146\text{ m}}{\left(700\frac{\text{W}}{v}\right)(0.0127\text{ m})}$$

$$\frac{1}{U_i} = 0.00336\frac{\text{m}^2\cdot\text{K}}{\text{W}}.$$

So,

$$U_i = 297.9\frac{\text{W}}{\text{m}^2\cdot\text{K}}.$$

The log mean temperature difference is calculated as

$$\Delta\bar{T}_{\text{ln}} = \frac{\Delta T_1 - \Delta T_2}{\ln\left(\frac{\Delta T_1}{\Delta T_2}\right)} = \frac{(50-4)-(90-30)^\circ\text{C}}{\ln\left(\frac{(50-4)}{(90-30)}\right)} = \frac{46-60^\circ\text{C}}{\ln\left(\frac{46}{60}\right)}$$

$$= \frac{-14^\circ\text{C}}{\ln(0.766)} = 52.7^\circ\text{C}.$$

The rate of heat transfer for the heat exchanger is then calculated from

$$q = U_i A_i \Delta \bar{T}_{\ln}. \tag{10.29}$$

Rearranging and substituting known values gives

$$A_i = \pi d_i L = \frac{q}{U_i \Delta \bar{T}_{\ln}} = \frac{9,924.2 \text{ W}}{\left(297.9 \frac{\text{W}}{\text{m}^2\text{K}}\right)(52.7 \text{ K})} = 0.632 \text{ m}^2.$$

Now,

$$L = \frac{A_i}{\pi d_i} = \frac{0.632 \text{ m}^2}{\pi(0.02291 \text{ m})} = 8.78 \text{ m}.$$

So, the heat exchanger must be 8.78 m long to provide sufficient heat transfer between these fluids.

6. *Check the solution*: The solution may be checked, as shown in previous problems, by using the overall heat-transfer coefficient based on the outside surface area.

7. *Generalize the solution strategy*: Heat-transfer rates in heat exchangers are dependent on the flow conditions within the exchanger, as well as the temperatures of the two media exchanging heat. How will the required length change if the flow rate of the milk is increased? Note that this will increase the heat requirements based on the enthalpy balance, but the convective heat-transfer coefficient would also increase.

Practice problem 10.15

Repeat problem 10.14 above for cocurrent operation.

10.2   Unsteady-state heat transfer

The previous problems have dealt with steady-state heat transfer, where temperatures at any point do not change with time. However, there are numerous instances in the food industry where unsteady-state heat transfer is important. One extremely important food process related to unsteady-state heat transfer is conventional thermal processing, where a can of food is heated in a retort to a temperature sufficient to destroy the microorganisms. The rate of temperature increase (unsteady-state heat transfer) determines the time needed to process the food. Food technologists have developed methods to calculate process time in a retort based on empirical measures, but the principles that underlie these calculations are based on unsteady-state heat transfer. Other examples where unsteady-state heat-transfer principles must be applied in food processing include heating water or a food product in a pot or batch kettle, cooking food in an oven, and chilling of fruits and vegetables.

The unsteady-state heat-transfer equation for an infinite slab, a solid object in one dimension, x, assuming constant physical properties, is given as

$$\frac{\partial T}{\partial t} = \frac{k}{\rho c_p} \frac{\partial T^2}{\partial^2 x}.$$ (10.31)

Solution to this partial differential equation is generally given as a series solution for T in terms of time and spatial dimension, x, and depends on the boundary conditions that apply to the particular problem to be solved.

The conditions at the surface of the material being heated or cooled dictate the method of solution of equation (10.31). The relative importance of convection at the surface to conduction within the object is determined by calculation of the Biot number,

$$N_{Bi} = \frac{hD}{k}$$ (10.32)

where h is the convective heat-transfer coefficient for the fluid flowing across the surface of the object, D is some characteristic dimension of the object, and k is the thermal conductivity of the solid object. If N_{Bi} is less than 0.1, k is large compared with hD, meaning that conduction in the solid is rapid compared with convection at the surface; thus, convection of heat to the surface governs the rate of heating or cooling. Heating of metal objects falls into this category. If N_{Bi} is greater than 40, hD is much greater than k, meaning that surface convection is rapid compared with conduction within the solid; thus, conduction within the object governs the rate of heating or cooling. Steam condensing on an object falls into this category. If N_{Bi} falls between 0.1 and 40, both conduction and convection heat transfer are important. Cooling an apple in a refrigerator is an example of when both heat transfer mechanisms are important.

The solution to equation (10.31) when N_{Bi} is greater than 0.1 is generally provided in graphical form (i.e., Heisler or Gurney–Lurie charts). Graphical solutions are available for three standard geometries: infinite slab, infinite cylinder, and sphere. If the object undergoing unsteady-state heat transfer fits one of these three standard geometries, then these charts can be used directly to relate temperature at any time at specific points within the object (e.g., center point). In these charts, the unaccomplished temperature difference, given as

$$Y = \frac{(T_a - T)}{(T_a - T_i)}$$ (10.33)

is plotted against the Fourier number,

$$N_{Fo} = \frac{k}{\rho c_p} \frac{t}{D^2}$$ (10.34)

where T_a is the ambient temperature, T_i is the initial temperature of the object, k is the thermal conductivity of the object, ρ is the density of the object, c_p is the specific heat of the object, and D is a characteristic dimension of the object (half thickness of an infinite slab, radius of an infinite cylinder, or radius of a sphere).

When N_{Bi} is less than 0.1, the temperature of the solid object is considered to be uniform throughout because conduction heat transfer is quite rapid relative to convection at the surface. Thus, a heat balance between the rate of heat transfer and the heat energy transferred can be written as

$$\dot{q} = \frac{dq}{dt} = mc_p \frac{dT}{dt} = hA\Delta T \tag{10.35}$$

where ΔT is the temperature difference between the fluid and the object, which changes with time as heating or cooling occurs. Equation (10.35) can be solved by separation of variables followed by integration over the proper boundary conditions. At $t = 0$, the initial ΔT is governed by the initial conditions $(T_a - T_i)$, where T_a is the ambient temperature condition of the fluid surrounding the solid object, and T_i is the initial temperature of the object. Over time, the temperature of the object changes until eventually the object attains the temperature of the ambient condition.

Solving equation (10.35) by separation of variables (see section 4.3) gives

$$mc_p \int_{T_i}^{T} \frac{dT}{(T_a - T)} = hA \int_0^t dt \tag{10.36}$$

and

$$\ln\left(\frac{T_a - T}{T_a - T_i}\right) = -\frac{hA}{mc_p}t$$

and

$$\left(\frac{T_a - T}{T_a - T_i}\right) = \exp\left\{-\frac{hA}{mc_p}t\right\}. \tag{10.37}$$

Equation (10.37) provides a relationship between temperature and time for heating or cooling of an object where the internal temperature is uniform. This applies for conditions when the convective heat-transfer coefficient is very low (natural convection) and the thermal conductivity of the object is high (good heat conductor). Metal objects heating or cooling in still air typically fall into this category. Another application of equation (10.37) is for heating or cooling of a fluid in well-agitated vessels, as might occur during heating of a fluid food in a stirred kettle. In this case, the temperature within the kettle can be considered approximately the same at any point, and convection heat transfer is determined by heating from the steam.

Problems 10.16 and 10.17 provide examples of unsteady-state heat transfer as applied in the food processing industry.

Worked problem 10.16

Find how long it would take to cool the center of an apple (10 cm in diameter) from 25°C to 10°C when submerged in cooling water at 1°C, where $h_o = 200$ W/m²·K. For apples, $k = 0.355$ W/m·K; $c_p = 3.6$ kJ/kg·K; and $\rho = 870$ kg/m³.

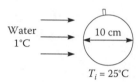

$$T_i = 25°C$$

Figure 10.8 Cooling of an apple

Solution:

1. *Motivation*: This is a problem dealing with unsteady-state heat transfer. Many startup or batch processes involve unsteady-state heating or cooling. This is an extremely important concept for the food scientist to understand, since the principles of unsteady-state heat transfer are required to calculate sterilization times to ensure food safety.

2. *Define the problem*: An apple is submerged in cool water (Figure 10.8). The initial temperature and final temperature of the apple are given, as well as the water temperature. The convective heat-transfer coefficient and thermal conductivity are also given, along with the density and specific heat of the apple. We are asked to find the time required for the apple to reach the final temperature.

3. *Think about the problem*: Solution of the unsteady-state heat-transfer problem depends on the Biot number (hD/k) used in conjunction with either the negligible internal resistance equation or with the unsteady-state heat-transfer charts. After determining which method to use, the cooling time may be calculated.

4. *Plan the solution*: We first must calculate the Biot number and then calculate the time required for the apple to cool. For an apple, we must assume spherical geometry.

5. *Implement the solution strategy*: The equation for the Biot number is

$$N_{Bi} = \frac{hD}{k}. \tag{10.32}$$

If the Biot number is greater than 0.1, the chart method must be used to solve for the time. The known values may be substituted into equation (10.32) to give

$$N_{Bi} = \frac{\left(200\,\frac{W}{m^2 \cdot K}\right)(0.05\text{ m})}{\left(0.355\,\frac{W}{m \cdot K}\right)} = 28.17.$$

The chart method must be used. The nondimensional temperature difference must be calculated, since all of the components for it have been given.

$$\frac{T_a - T}{T_a - T_i} = \frac{1 - 10}{1 - 25} = 0.375$$

$$m = \frac{1}{N_{Bi}} = 0.035$$

Using these two values and the appropriate chart for spherical geometry, the Fourier number is found to be approximately 0.25. The Fourier number is written as

$$N_{Fo} = \frac{k}{\rho c_p}\frac{t}{D^2} \tag{10.34}$$

where D is the diameter of the sphere.
Rewriting and substituting the given values gives

$$t = \frac{N_{Fo}\rho c_p d^2}{k} = \frac{(0.2)\left(870\,\frac{kg}{m^3}\right)\left(3600\,\frac{J}{kg \cdot K}\right)(0.05\text{ m})^2}{0.355\,\frac{W}{m \cdot K}} = 4{,}411\text{ s} = 1.23\text{ h}.$$

It will take over 6 hours for this apple to cool from 25°C to 10°C in this water.

6. *Check the solution*: Does this seem like a reasonable time? In your experience, will it take about this long for the center temperature of the apple to decrease from 25°C to 10°C? Unfortunately, there is no way to independently verify this calculation except by actually performing an experiment.

7. *Generalize the solution strategy*: The method allows one to analyze what properties of the system influence the time required to cool the apple. Temperatures, diameter, and the convective heat-transfer coefficient are among the important parameters. If you wanted to cool the apple as rapidly as possible, what combination of conditions would you use?

Practice problem 10.17

One way to measure the convective heat-transfer coefficient around a food product is to heat a metal piece of the same shape under the same conditions and measure the temperature of the piece as a function of time. For the previous problem, we might use a 10-cm copper ball ($\rho = 8954$ kg/m³, $c_p = 3831$ kg·K,

$k = 388$ W/m · K). If the copper ball started at 25°C and cooled to 10°C in 45 min in cooling water at 1°C, what is the convective heat transfer coefficient, h?

10.3 Freezing

Another example of unsteady-state heat transfer occurs during freezing of foods. Freezing is a very common unit operation in the food industry, as foods can be preserved for extended time periods in the frozen state. Growth of microorganisms and rates of other chemical reactions occur very slowly, if at all, at these low temperatures.

Although freezing is unsteady-state heat transfer, the equations derived previously cannot be used in this case. Upon freezing, there is a substantial release of latent heat of ice formation that is not accounted for in the simple heat-transfer equations used previously. Formally, equation (10.31) can be modified through addition of an energy-generation term or by accounting for the energy contribution of latent heat generation in the specific-heat term. However, a simple equation has been developed by Planck that accounts for the phase change associated with freezing.

Planck's model assumes that the object to be frozen is already at its freezing temperature, T_f, and that freezing occurs due to exposure to cold air at T_a. The model also assumes that all freezing takes place at T_f, no cooling below T_f occurs, and the physical properties of the material are constant. Despite these assumptions, Planck's equation gives a reasonable estimate of freezing times in food processing.

The derivation of Planck's equation depends on the particular geometry of the food, but the general form of the final equation is the same. For a slab of material being frozen from both sides, Planck assumed that there was a balance between the rate of heat transfer out of the freezing slab and the rate of heat generation by the phase change. Layers of frozen material form at the surface and work their way in to the center of the slab, as seen in Figure 10.9.

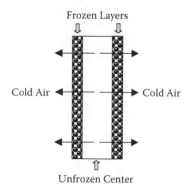

Figure 10.9 Freezing of a slab of food

When the freezing interfaces from each side meet at the center, the object is considered frozen. The rate of heat transfer out of the freezing interface is dependent on the rate of conduction through the frozen layer and convection into the ambient air. The rate of heat generation is the product of the rate of ice formation and the latent heat of crystallization. By assuming that the rate of heat generation due to ice formation is balanced by the rate of heat removal, Planck developed an equation for predicting freezing time. For an infinite slab (as in Figure 10.9), Planck's equation for freezing time, t_f, is given as

$$t_f = \frac{\rho_f \lambda_f}{(T_f - T_a)} \left\{ \frac{a^2}{8k_f} + \frac{1}{2h} \right\}$$

(10.38)

where ρ_f is the density of the frozen food, k_f is the thermal conductivity of the frozen food, λ_f is the latent heat of formation, T_f is the initial freezing temperature, T_a is the ambient temperature in the freezer, and h is the convective heat-transfer coefficient between the air and the food. The dimension, a, in the case of a slab being frozen from both sides is the thickness of the slab.

Similar equations can be derived for other geometries, so that Planck's equation is often given in general form,

$$t_f = \frac{\rho_f \lambda_f}{(T_f - T_a)} \left\{ \frac{Pa^2}{k_f} + \frac{Ra}{h} \right\}$$

(10.39)

where the coefficients P and R can be found for several different geometries according to Table 10.1.

Application of Planck's equation is demonstrated in problems 10.18 and 10.19.

Worked problem 10.18

How long would it take to freeze an apple (10-cm diameter) if cold air at $-30°C$ blows across it (Figure 10.10) to give a convective heat-transfer coefficient of 100 W/m²·K? For unfrozen apples: $\rho_u = 870$ kg/m³, $c_{pu} = 3.6$ kJ/kg·K, and $k_u = 0.355$ W/m·K; for frozen apples: $\rho_f = 760$ kg/m³, $c_{pf} = 2.2$ kJ/kg·K, and

Table 10.1 Values of Constants P and R in Planck's Equation (Equation 10.29) for Different Geometries

Geometry	P	R	a
Infinite slab	1/2	1/8	thickness
Infinite cylinder	1/4	1/16	diameter
Sphere	1/6	1/24	diameter
Cube	1/8	1/24	edge length

$k_f = 1.10$ W/m·K. Use a moisture content of 0.88 and a freezing temperature of $T_f = -0.9°C$. Assume that the apple is initially at its freezing point.

Solution:

1. *Motivation:* Food freezing is a complicated heat-transfer operation. It involves both a change in temperature and a change in state (water to ice). Many food process operations involve freezing as a means of preservation.
2. *Define the problem:* The thermal properties of frozen and unfrozen apples are given, as well as the freezing temperature of the apple. The convective heat-transfer coefficient as air blows across the apple is also given. We are asked to find the time required to freeze the apple.
3. *Think about the problem:* The latent heat of freezing is dependent on the moisture content (0.88) of the apple. Several methods have been used to approximate the latent heat generated during freezing of foods. We will use a very simple approximation based on the product of moisture content and latent heat of pure water. We will also assume spherical geometry (Figure 10.10).
4. *Plan the solution:* We will first calculate the latent heat of freezing for the water in the apple. Afterwards, we will calculate the freezing time with the Planck equation written for spherical geometry. This equation assumes the apple starts at its freezing temperature (−0.9°C) and is frozen once the ice front reaches the center of the sphere.
5. *Implement the solution strategy:* The latent heat of freezing is estimated as

$$\lambda_f = MC \cdot \lambda_{water} = (0.88)\left(333.2 \frac{kJ}{kg}\right) = 293.2 \frac{kJ}{kg}.$$

Planck's equation for freezing of a sphere is

$$t_f = \frac{\rho_f \lambda_f}{(T_f - T_a)}\left[\frac{1}{24}\frac{a^2}{k_f} + \frac{1}{6}\frac{a}{h}\right]$$ (10.40)

where a is the diameter of the apple and T_a is the ambient temperature in the freezer.

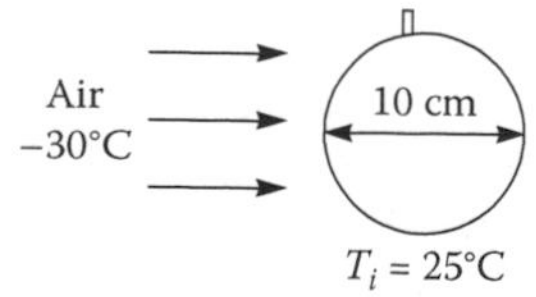

Air
−30°C

10 cm

$T_i = 25°C$

Figure 10.10 Freezing of an apple

Substituting the known quantities and then solving gives

$$t_f = \frac{\left(760\,\frac{kg}{m^3}\right)\left(293.2(10^3)\,\frac{J}{kg}\right)}{(-0.9+30^\circ C)}\left[\frac{1}{24}\frac{(0.1\,m)^2}{1.10\,\frac{W}{m\cdot K}} + \frac{1}{6}\frac{(0.1\,m)}{100\,\frac{W}{m^2\cdot K}}\right]$$

$$= 4,176.3\,s = 1.16\,h.$$

6. *Check the solution*: Does this sound reasonable? Will it take a little over an hour to freeze an apple solid in this cold freezer?
7. *Generalize the solution strategy*: In equation (10.15), the effect of several parameters on freezing time can be seen. Which combination of conditions would you choose to promote the most rapid freezing process?

Practice problem 10.19

The fan in the freezer in problem 10.18 has broken down, so that there is no air flow across the apple. We can estimate that $h_o = 5\,W/m^2\cdot K$ under these conditions. If we need to move the apples from the freezer after 2 hours, will they be frozen?

10.4   Radiation heat transfer

The final mode of heat transfer to consider is radiation. All objects give off electromagnetic radiation, depending on their temperature. These electromagnetic waves pass through the environment surrounding a hot object. When that radiation impinges upon a colder body, some of the radiation is absorbed by the colder body. The temperature of the colder body increases according to the amount of energy absorbed.

The heat flow by radiation from a hot object is calculated as

$$\dot{q} = \varepsilon A \sigma T^4 \tag{10.41}$$

where

      ε = emissivity of the object
      A = surface area from which heat is transferred
      σ = Stefan-Boltzmann constant ($5.669 \times 10^{-8}\,W/m^2\cdot K^4$)
      T = absolute temperature (K).

The emissivity of an object depends upon its inherent characteristics. For a blackbody, defined as an object that absorbs all radiation energy that impacts upon it (reflecting none), emissivity is 1. Virtually all materials have emissivity less than 1, since they are not perfect blackbodies. Furthermore, according to Kirchoff's law, emissivity and absorptivity (the ability of a material to absorb radiation) are equal.

The amount of radiation energy passed between two objects depends on numerous factors, including the emissivity of the hotter object, the absorptivity of the colder object, the size and shape of the two objects, the manner in which the two objects are situated near each other, and the temperatures of the two objects. The view factor, a measure of how much radiation from the hotter object impinges on the colder object, depends on the physical arrangement of the two objects. Calculation of the rate of heat transfer between two objects often comes down to determining the correct view factor. A general equation for the rate of heat transfer by radiation between two objects is given as

$$\dot{q} = \sigma A_1 \xi_{12} \left(T_1^4 - T_2^4 \right)$$

(10.42)

where ξ_{12} is the view factor between the two objects, and T_1 and T_2 are absolute temperatures of the hotter and colder objects, respectively.

Worked problem 10.20

A loaf of bread (surface area of 0.15 m², $\varepsilon = 0.8$) at a temperature of 100°C is in an oven at 350°C (Figure 10.11). Assuming the oven acts like a blackbody radiator ($\varepsilon = 1$), find the radiative heat gain of the loaf of bread, where $\sigma = 5.676 \times 10^{-8}$ (W/m²·K⁴).

Solution:

1. *Motivation*: Radiation is often overlooked as a mode of heat transport. It does, however, affect certain food processes. The food scientist should be familiar with the concept of radiation.
2. *Define the problem*: We are given the bread temperature, surface area, and emissivity as well as the oven characteristics and asked to calculate the rate of heat transfer.
3. *Think about the problem*: Because the oven completely surrounds the bread, the view factor is equivalent to the emissivity of the bread. The rate of heat transfer may then be calculated with the radiation equation.

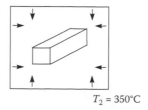

$T_2 = 350°C$

Figwure 10.11 Radiative heat transfer in an oven

4. *Plan the solution*: We will substitute the known values into the radiation equation and solve for the rate of heat transfer.
5. *Implement the solution strategy*: The radiative heat-transfer equation is

$$q = A_1 \xi_{12} \left(T_1^4 - T_2^4 \right) = \sigma A_1 \varepsilon_1 \left(T_1^4 - T_2^4 \right). \tag{10.43}$$

Here, $\xi_{12} = \varepsilon_1$ for the case of a small object completely surrounded by a large object.
Substituting the known values and solving gives

$$q = \left[5.676(10^{-8}) \, \text{W} \middle/ \text{m}^2 \cdot \text{K}^4 \right] (0.15 \, \text{m}^2)(0.8) \left[(373 \, \text{K})^4 - (623 \, \text{K})^4 \right]$$

$$= -894.2 \, \text{W}.$$

So, 894.2 W is transferred from the oven to the bread under these conditions.

6. *Check the solution*: No independent method for checking the calculation is available.
7. *Generalize the solution strategy*: This radiation example consisted of a blackbody emission source. How would it change for an oven that was not a blackbody?

Practice problem 10.21

A conveyor filled with molded chocolate bars at 30°C passes through a cooling tunnel with black surfaces kept at 10°C to provide radiative cooling. Assuming the cooling-tunnel surface acts as a black body radiator ($\varepsilon = 1$), while the chocolate has an emissivity of 0.90. Calculate the rate of radiative heat transfer per unit area from the chocolate to the cooling tunnel.

chapter eleven

Mass transfer

Mass transfer is an important part of many food processing operations, including transfer of water to air during drying and humidification, extraction of solutes during separation processes and leaching, provision of dissolved oxygen during aerobic fermentation, and delivery of solutes during crystallization. It involves the migration of a component of a mixture or food product due to chemical potential differences that are typically produced by nonuniform concentration. The presence of a concentration gradient within a food material leads to molecular diffusion of the component from areas of high concentration to those with low concentration. For example, if you put a cucumber in a salt solution to pickle, the salt will diffuse into the cucumber because the concentration of salt in the liquid phase of the cucumber is lower than the concentration of salt in the solution. This unsteady-state mass transfer will continue until the salt concentration everywhere inside the cucumber is in equilibrium with the salt concentration in the solution outside the cucumber. An example of steady-state mass transfer is the transfer of water vapor released during respiration through packaging films covering fresh fruits and vegetables. With liquids and gases, an enhancement of molecular diffusion, called convective mass transfer, occurs when there is flow or agitation that speeds up the rate of mass transfer, which is analogous to what occurs during convective heat transfer.

Water is the component that we are the most frequently interested in during food processing and storage. Moisture concentration differences can develop simply due to changes in the weather that affect the temperature or humidity levels of air. In addition, changes in temperature during food processing, such as heating of air in an oven or cooling of air in a refrigerator, also lead to significant moisture gradients that drive mass transfer of water to or from food. Psychrometrics is the field of engineering that includes the determination of the physical and thermodynamic properties of moist air. With these properties, it is possible to predict the migration of moisture in food products during heating, drying, and storage.

11.1 Psychrometrics

Psychrometrics or psychrometry are engineering terms that refer to the set of principles used in the determination of physical and thermodynamic properties of any system consisting of a gas–vapor mixture such as those described in chapter 6. However, psychrometrics is most commonly used to describe determination of the properties of mixtures of water vapor and air because of their importance in heating, ventilating, and air-conditioning, and in meteorology. There are two major approaches to determining the properties of air–water mixtures: calculation of the properties based on the thermodynamics of the mixture and the gas laws, or the use of the psychrometric chart. The heart of psychrometrics is the measurement of the dry-bulb and wet-bulb temperatures. The dry-bulb temperature (T) is the temperature you get using an ordinary thermometer. The wet-bulb temperature (T_{wb}) is read by a temperature sensor covered with a wet wick and exposed to relatively fast moving (>5 m/s) unsaturated air. This provides the temperature where the rate of convective heat transfer is equal to the rate of vaporization. In the case of air–water mixtures at atmospheric pressure, the wet-bulb temperature is also equal to the thermodynamic wet-bulb temperature reached by unsaturated air when it is adiabatically (no heat loss or gain) saturated by evaporating water. This allows us to use thermodynamic principles to determine the properties of moist air. Therefore, these two temperatures allow us to determine all of the other important properties of moist air. An instrument that measures these two temperatures is called a psychrometer.

11.1.1 Moist-air properties using thermodynamic principles and the gas laws

Table 11.1 gives the relationships for calculating the important properties of moist air. These properties are based on the gas laws presented in chapter

Table 11.1 Equations for Calculating the Properties of Moist Air when Given the Wet-Bulb Temperature (T_{wb}) and the Dry-Bulb Temperature (T) in °C

Property (units)	Description	Equation
P_{atm} (kPa)	Atmospheric pressure	$P_{atm} \cong 101.325$ at sea level
p^o_{wt} (kPa)	Saturation vapor pressure at the wet-bulb temperature	Saturation vapor pressure read from steam table at T_{wb}
p_w (kPa)	Water vapor partial pressure	$p_w = p^o_{wb} - \dfrac{(P_{atm} - p^o_{wb})(T - T_{wb})}{1555.56 - 0.722 T_{wb}}$
p_a (kPa)	Dry air partial pressure	$p_a = P_{atm} - p_w = 101.325 - p_w$
T_{dp} (°C)	Dew-point temperature	p_w @ T $= p_w$ @ T_{dp} on steam table
W (kg H$_2$O/kg dry air)	Specific humidity ratio (dry basis moisture content)	$W = 0.622 \dfrac{p_w}{p_a} = 0.622 \dfrac{p_w}{P_{atm} - p_w}$
RH (%)	Relative humidity	$\%RH \dfrac{p_w}{p^o_w} \times 100\%$
V'_m (m³/kg dry air)	Specific volume of moist air	$V'_m = (0.082T + 22.4)\left(\dfrac{1}{29} + \dfrac{W}{18}\right)$
$\Delta H'_m$ (kJ/kg dry air)	Specific enthalpy of moist air when $T_o = 0$°C	$\Delta H'_m = \Delta H'_a + W\Delta H'_w = (1.005 + 1.88W)(T_a - T_0) + 2501.4W$

6 and thermodynamic principles such as those used in chapter 8. These equations, along with use of the steam tables to find the saturation vapor pressure (p°_{w}) at a given temperature, as described in chapter 8, allow the calculation or determination of all other moist-air properties when the wet-bulb temperature (T_{wb}) and the dry-bulb temperature (T) are known. This includes the relative humidity (% RH), which is the ratio of the actual water vapor pressure in the air to the value at saturation at a given temperature. It provides a measure of the percentage of the water-carrying capacity of the air being utilized. In contrast, the actual amount of water that is in the air is described by the specific humidity ratio (W), which is the dry-air-basis moisture content.

As seen in Table 11.1, the specific enthalpy of moist air $(\Delta H'_{m})$ is the sum of the specific enthalpy content of the dry air $(\Delta H'_{a})$ and the specific enthalpy content of water vapor $(\Delta H'_{w})$ multiplied by the moisture content of the moist air (W). The specific enthalpy of moist air can also be found from the average specific heat capacity values of water vapor and dry air along with the latent heat of vaporization of water at a reference temperature of 0°C (Table 11.1).

The dew-point temperature (T_{dp}) is frequently referred to in weather reports. It is the temperature when unsaturated, moist air that has been cooled at a constant pressure and moisture content becomes saturated and begins to condense. This frequently happens when the air cools down rapidly after the sun goes down at the constant atmospheric pressure. When you go outside in the morning and the grass is covered with dew or frost, it means that moisture has condensed out of the air because the air temperature went below the dew-point temperature. The dew-point temperature can be determined using the steam tables by finding the temperature where the water vapor partial pressure in the air becomes equal to the saturation vapor pressure $(p_{w}$ at $T = p^{\circ}_{w}$ at $T_{dp})$. Likewise, when the dew-point temperature is known, the steam tables can be used to find the water vapor partial pressure in the moist air.

11.1.2   *Moist-air properties on psychrometric chart*

The moist-air properties that can be calculated from the equations in Table 11.1 with the use of the steam tables have been collected in the psychrometric chart.[1] Because of the volume of information contained on the psychrometric chart,[1] it is confusing when you first look at it. Therefore, the various lines and curves are represented schematically in Figures 11.1 and 11.2 to help you sort them out. The moist-air properties with typical units and their locations on the chart are also summarized in Table 11.2. Note that the specific humidity ratio (W) is frequently shown in units of g H_2O/kg dry air.

---

[1] Steam tables and psychrometric units are widely available online at the Engineering Toolbox (http://www.engineeringtoolbox.com) or engineering handbook including *Perry's Chemical Engineer's Handbook.*

Table 11.2 Location of Moist-Air Properties on the Psychrometric Chart

Property (units)	Description	Location on Chart
W (g H_2O/kg dry air)	Specific humidity ratio (dry basis moisture content)	*y-axis*
RH (%)	Relative humidity	Main curves with the outer curve at 100% saturation
T (°C)	Dry-bulb temperature	*x-axis*
T_{wb} (°C)	Wet-bulb temperature	Temperature where constant enthalpy lines cross saturation curve
T_{dp} (°C)	Dew-point temperature	Temperature where constant W lines cross the saturation curve
V'_m (m³/kg dry air)	Specific volume of moist air	Vertical diagonal lines
$\Delta H'_{ms}$ (kJ/kg dry air)	Specific enthalpy of saturated air	Horizontal diagonal lines
D (kJ/kg dry air)	Enthalpy deviation from saturation	Vertical curves (negative)

The main curves shown in both figures are the constant relative humidity (% RH) lines, with the outer curve representing saturation (100% RH). The lines coming from the *x*-axis shown in Figure 11.1 are the constant dry-bulb temperature (T) lines. The *y*-axis shows the levels of constant dry-basis moisture content (W), which is also known as the humidity ratio or the specific humidity. As shown in Figure 11.1, the dew-point temperature (T_{dp}) for any point on the chart can be found by following the constant-humidity ratio (W) line to where it crosses the saturation curve and reading the temperature at that point off the *x*-axis.

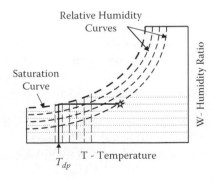

Figure 11.1 Determination of the dew-point temperature (T_{dp}) on the psychrometric chart

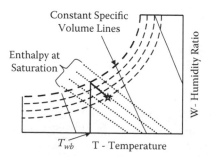

Figure 11.2 Determination of the wet-bulb temperature (T_{wb}) on the psychrometric chart

Figure 11.2 shows the constant enthalpy at saturation ($\Delta H'_{ms}$) lines and the constant specific volume (V'_m) lines. Additional vertically diagonal curves on the psychrometric chart (not shown) give the deviation (D) of the enthalpy from saturation at unsaturated conditions. Then, the enthalpy for unsaturated conditions ($\Delta H'_m$) is found by decreasing the value of enthalpy at saturation by the enthalpy deviation ($\Delta H'_m = \Delta H'_{ms} + D$). As shown in Figure 11.2, the wet-bulb temperature (T_{wb}) for a given point is found by following the constant enthalpy at saturation lines to the saturation curve and reading the temperature for that point from the x-axis.

11.1.3 Using the psychrometric chart to find the state of air during processing

In many food processes, the condition of air changes as part of the process, including air-conditioning, heating, drying, evaporative cooling, and humidification and dehumidification of air. The psychrometric chart can be used to follow the state of the air in these types of processes, when the state of the air at specific points of the process is known. For example, water is neither added nor removed during heating of air. Therefore, if the state of incoming air is known, the state of the air after heating can be found on the chart by following the constant-moisture-content line to the final air temperature. Likewise, during cooling of air, the constant-moisture-content line is followed, but only until it reaches the saturation curve. Then the saturation curve is followed to the final air temperature.

During drying, evaporative cooling, and humidification, the heat of evaporation required is supplied only by the air, which is called an adiabatic saturation process, where the total enthalpy remains constant. Therefore, if the state of the air entering the process is known, the final state of the air can be determined by following the constant-enthalpy lines if given one piece of information on the final state of the air, such as the dry-bulb temperature or relative humidity.

Worked problem 11.1

Air at 25°C and 70% RH is heated to 100°C. It is then introduced into a spray drier. Exit temperature from the drier is 80°C. Determine the following:

- The relative humidity (% RH) of the heated air
- Wet-bulb temperature (T_{wb}) of heated air
- The relative humidity (% RH) of air leaving the drier
- Moisture content (W) of air leaving drier
- Dew-point temperature (T_{dp}) of air leaving drier

Solution:

1. *Motivation*: This problem involves simple heating and humidification of air. These are common operations in drying processes.
2. *Define the problem*: We are given the initial conditions of the air, as well its intermediate and final temperatures, and asked to determine the properties of air at the different conditions.
3. *Think about the problem*: During simple heating, the specific humidity (total amount of water in the air) does not change, and during spray drying, we assume the process occurs at constant enthalpy. The psychrometric chart allows one to quickly determine air properties when two of these properties are known.
4. *Plan the solution*:
   a. The specific humidity ratio does not change during the heating stage. The relative humidity of the hot air may be determined by visually interpolating between the lines of constant relative humidity that surround the new point on the psychrometric chart.
   b. The T_{wb} may be determined by following the constant T_{wb} lines (essentially the constant enthalpy at saturation lines) to the saturation line from the intermediate point on the chart. Then, the value of T_{wb} is read off of the temperature axis.
   c. The final point of the process for air leaving the dryer is found by following the constant-enthalpy-at-saturation line from the intermediate point on the chart to the final-temperature line. We will follow the same procedure as in (a), using the final point on the chart.
   d. The specific humidity is found by moving straight across (horizontally) from the final point to the specific humidity ratio axis and recording the value.
   e. The T_{dp} is determined by moving horizontally to the left from the final point along the constant-specific-humidity-ratio line to intersect the saturation curve. The temperature for this point is the dew point.
5. *Implement the solution strategy*: The schematic of the psychrometric chart in Figure 11.3 shows what the processes look like.

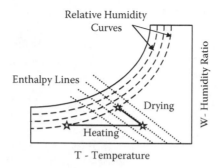

Figure 11.3 Schematic of heating and drying processes on the psychrometric chart

- Visually interpolating from the chart, the relative humidity (% RH) for the heated air is approximately 2.2%.
- The T_{wb} is near 37°C, from the chart.
- The relative humidity (% RH) of the air leaving the drier is near 7.5%.
- The moisture content of the air leaving the drier is approximately 0.023 kg water/kg dry air.
- The T_{dp} is about 27°C.

6. *Check the solution:* The solution can be checked with the thermodynamic equations that apply to air–water mixtures in Table 11.1. These equations are tedious at times, which is why the psychrometric chart was developed.
7. *Generalize the solution strategy:* The psychrometric chart was used to determine the properties of air and to analyze how these properties change under some simple processes. Given any two properties on the chart, the remaining five can be determined.

Practice problem 11.2

Caking of a dried formulated food powder was found to occur when exposed to air with a relative humidity in excess of 5%. In a spray drier, ambient air at 20°C and 50% RH is heated to 171°C and introduced into the drier. Calculate the minimum exit temperature for the air from the spray drier such that caking of the dried powder does not occur.

11.2 Molecular diffusion

When a concentration gradient exists, the driving force behind molecular diffusion is random molecular motion. Even though the molecules move randomly, there is a net movement of molecules away from areas of high concentration to those of low concentration until eventually the concentration is uniform throughout. The extent of random motion depends on the state of the material due to the differences in molecular mobility. Therefore, the rate of diffusion depends on the state of the materials, with gases diffusing faster than liquids and liquids diffusing faster than solids.

The mathematical relationship that is used to model molecular diffusion in the presence of a concentration gradient is Fick's law, which can be written in the case of one-dimensional mass transfer as

$$\frac{\dot{m}_A}{A} = -D_{AB}\frac{dC_A}{dx} \tag{11.1}$$

where $\dot{m}_A$ is the mass flow rate (kg/s), A is the surface area (m²), x is the distance that mass is transferring across (m), and D_{AB} is the mass diffusivity of component A into B with units of m²/s when the concentration (C_A) has units of (kg A)/m³. D_{AB} is a proportionality constant that depends on the molecular mobility of the substances in the system, with gases having a much higher diffusivity (~10^{-5} m²/s for binary gas mixtures) than liquids (~10^{-9} m²/s for aqueous solutions) or solids (~10^{-10} m²/s for water or solutes in solid food materials). It is analogous to the thermal diffusivity (α) in heat transfer.

In the case of steady-state diffusion of a gas or liquid A with a concentration gradient of ($C_{A1} - C_{A2}$) through a solid, B, that has a thickness of $x = x_2 - x_1$, we have the situation depicted in Figure 11.4. Separating the variables and integrating equation (11.1) using integration rule 1 from chapter 4, equation (4.17) gives

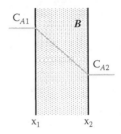

Figure 11.4 Steady-state, one-dimensional diffusion of gas or liquid A through rectangular solid B

$$\frac{\dot{m}_A}{A}\int_{x_1}^{x_2} dx = -D_{AB}\int_{C_{A1}}^{C_{A2}} dC_A \tag{11.2}$$

$$\frac{m_A}{A} = D_{AB}\frac{C_{A1}-C_{A2}}{x_2 - x_1}. \tag{11.3}$$

In the case of gases, frequently what is used is the partial pressure of the gas (p_A) at the surface rather than the concentration. The solid concentration at the surface can then be described using Hook's law:

$$C_A = Sp_A \tag{11.4}$$

where S is the solubility of the gas in the solid. Replacing the concentration then gives:

$$\frac{m_A}{A} = D_{AB}S\frac{p_{A1}-p_{A2}}{x_2 - x_1}. \tag{11.5}$$

The quantity $D_{AB}S$ is known as the permeability ($\mathcal{P}$). The typical units used for S and $\mathcal{P}$ vary widely, so care must be taken to properly match the units of all quantities when getting these values from reference texts.

Worked problem 11.3

A 3-cm × 3-cm window of 0.1-mm-thick cellophane is found on a box of pasta. If the 25°C air outside the box is at an average 80% RH and the inside air is initially at 5% RH, estimate the maximum moisture diffusion through the window per week (g/week). The permeability ($\mathcal{P}$) of water vapor in cellophane is 1.425×10^{-10} (cm³ @ STP·cm)/(cm²·s·Pa), and the saturation vapor pressure (p°_w) at 25°C is 3.179 kPa.

1. *Motivation*: This problem involves moisture transfer across packaging material. The ability of packaging material to serve as a moisture barrier is important in increasing the shelf life of dry food materials.
2. *Define the problem*: We are given the average conditions of the outside air and the initial condition of the inside air, as well the permeability of the cellophane packaging material, and are asked to determine the maximum amount of moisture that will diffuse through the cellophane window per week.
3. *Think about the problem*: The maximum rate of moisture migration through the cellophane occurs at the initial condition of the inside air, because the rate will go down as moisture diffuses into the package and reduces the concentration gradient. Since the main resistance to diffusion of the water vapor is through the film, all other resistances are neglected, and Fick's law applies as stated in equation (11.3) for a slab. The concentration at the slab boundaries can be determined using the solubility of water vapor in the film and the partial pressure of water vapor in the air at the boundaries. The solubility units will have to be transformed to give the correct result.
4. *Plan the solution*: The partial pressure of the water vapor on the inside and the outside of the film can be determined from the definition of the relative humidity in Table 11.1 and the saturation vapor pressure at 25°C. The version of Fick's law in equation (11.4) relates the partial pressure difference of water vapor at the surfaces of the slab to water vapor mass transfer. The solubility and the diffusivity are accounted for by the film permeability given in the problem. The given units of the permeability will have to be transformed from volume of water vapor at STP to mass of water vapor using the gas law and the molecular weight of water.
5. *Implement the solution strategy*: The partial pressures of water vapor inside and outside the package found using the relative humidity

definition in Table 11.1 are

$$\%RH = \frac{p_w}{p_w^o} \times 100 \rightarrow p_w = \frac{\%RH}{100} \times p_w^o.$$

Inside the package:

$$p_{wi} = \frac{5\%}{100} \times 3179 \text{ Pa} = 159 \text{ Pa}$$

Outside the package:

$$p_{wo} = \frac{80\%}{100} \times 3179 \text{ Pa} = 2543 \text{ Pa}$$

Then, applying equation (11.4), we have

$$m_w = AD_{AB}S\frac{p_{wo} - p_{wi}}{x_2 - x_1} = (3 \text{ cm} \cdot 3 \text{ cm})\mathcal{P}\frac{(2543 \text{ Pa} - 159 \text{ Pa})}{0.01 \text{ cm}}.$$

The meaning of the units given for the permeability is

$$\frac{\text{cm}^3 \text{ @ STP} \cdot \text{cm}}{\text{cm}^2 \cdot \text{s} \cdot \text{Pa}}$$

$$= \frac{(\text{amount of gas vapor}) (\text{film thickness})}{(\text{mass transfer area}) (\text{mass transfer time}) (\text{pressure difference across the film})}.$$

Since we want the answer in grams/week, we need to transform the time from seconds to weeks (1 week = 604,800 seconds). We also need to transform the water vapor partial volume (cm³ @ STP) to grams of water using the ideal gas law, equation 6.6, and the molecular weight of water (M_w = 18 g/g mol), where

$$m_w = n_w M_W = \frac{Pv_w}{RT} M_W$$

$$= \frac{1 \text{ (atm)} \cdot v_w \text{ (cm}^3 \text{ @ STP)}}{82.057 \text{ (cm}^3 \cdot \text{atm/g mol} \cdot \text{K)} \cdot 273.16 \text{ (K)}} \cdot 18 \text{(g/g mol)}$$

$$m_w = 8.03 \times 10^{-4} \left(\frac{\text{g}}{\text{cm}^3 \text{ @ STP}} \right) \cdot v_w \text{ (cm}^3 \text{ @ STP)}$$

standard temperature and pressure (STP) is 0°C and 1 atm of pressure. The gas constant value was found in Table 6.1. Therefore, the permeability in the correct units is

$$P = 1.425 \times 10^{-10} \frac{cm^3 @ STP \cdot cm}{cm^2 \cdot s \cdot Pa} \cdot 8.03$$

$$\times 10^{-4} \frac{gm}{cm^3 @ STP} \cdot 6.048 \times 10^5 \frac{s}{week}$$

$$P = 6.921 \times 10^{-8} \frac{g \cdot cm}{cm^2 \cdot week \cdot Pa}.$$

Then the final answer is

$$m_w = (3 \text{ cm} \cdot 3 \text{ cm})\left(6.921 \times 10^{-8} \frac{g \cdot cm}{cm^2 \cdot week \cdot Pa}\right)\frac{(2542 \text{ Pa} - 159 \text{ Pa})}{0.01 \text{ cm}}$$

$$= 0.15 \text{ (g/week)}.$$

6. *Check the solution:* According to this calculation, 0.15 g of water will diffuse into the pasta package each week through the cellophane window (neglecting water ingress through the cardboard package and its seals). No independent verification of this calculation is possible; however, from an order-of-magnitude standpoint, we would not expect a lot of water would transport into the package, based on experience, and 0.15 g of water is a small amount.

7. *Generalize the solution strategy:* This approach can be used to determine the steady-state mass transfer of any gas or vapor through a packaging film, where the film surface area and permeability, as well as the partial pressures or concentrations of the gas or vapor on both sides of the film, are known. More generally, this approach can be applied with the appropriate concentration units to any situation of one-dimensional steady-state mass transfer through a flat-plate solid, where the surface concentrations are known or can be determined from known information.

Practice problem 11.4

Plastic wrap (poly[vinylidene chloride]) is used to cover an 8.25-in.-diam. impermeable glass bowl containing lettuce. The 4°C refrigerated air outside the box is at an average 90% RH, and the lettuce respiration maintains the air inside the bowl at 100% RH. Assuming a good edge seal, estimate the amount of moisture that diffuses out of the container through the 0.1-mm-thick plastic cover per day (mg/day). The permeability (P) of water vapor in poly(vinylidene chloride) is 0.5×10^{-10} (cm^3 @ STP·cm)/(cm^2·s·cm Hg), and the saturation vapor pressure (p°_w) at 4°C is 875.8 Pa.

11.3   Convective mass transfer

When mass transfers across an interface where one or both of the phases
is a liquid, a gas, or a porous solid, diffusion is frequently enhanced by fluid
motion. This enhancement of diffusion,
called convective mass transfer, occurs
in a similar fashion to convective heat
transfer. In the case of mass transfer
into a bulk liquid or gas from a liquid
or solid interface, there is a boundary
layer with a thickness, z_g, that has a
concentration different from that of
the bulk, as shown in Figure 11.5. The

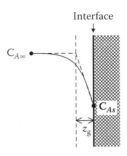

Figure 11.5 Steady-state, one-dimensional
convective mass transfer of gas or liquid
A across a fluid–solid interface

boundary layer thickness is usually unknown and depends on the fluid motion.
Therefore, the mass transfer by convection into a bulk fluid with an A concentration of C_A from a surface with an A concentration of C_A is described as

$$\frac{\dot{m}_A}{A} = k_m (C_{As} - C_{A\infty}) \tag{11.6}$$

where k_m is the convective mass-transfer coefficient. This coefficient lumps together
the molecular diffusivity and the unknown boundary layer thickness. It can be
determined experimentally or estimated using empirical correlation relationships
in the same manner as the convective heat-transfer coefficient.

The concentration described above is in units of (kg A/m³), which is the density of A in B. When working with gases and vapors, this can be converted to
partial pressures using the gas law principles described in chapter 6:

$$C_A = \rho_A = \frac{m}{V} = \frac{n_A M_A}{V} = \frac{p_A M_A}{RT}. \tag{11.7}$$

Therefore, for gases and vapors, we can calculate convective mass transfer in
terms of the partial pressure with the expression:

$$\frac{\dot{m}_A}{A} = \frac{k_m M_A}{RT} (p_{As} - p_{A\infty}). \tag{11.8}$$

As with convective heat transfer (see section 10.1.3), the empirical correlation relationships for convective mass transfer are also based on dimensionless numbers: in this case, the Sherwood number ($k_m D/D_{AB}$), the Reynolds
number ($\rho v D/\mu$), and the Schmidt number ($\mu/\rho D_{AB}$). The relationships take
the general form:

$$N_{Sh} = a + b(N_{Re})^c (N_{Sc})^e \tag{11.9}$$

where *a*, *b*, *c*, and *e* are the best-fit parameters for the model from data collected in a wide range of experiments for a particular geometry with a characteristic dimension, *D*. As is the case with the convective heat-transfer coefficient, empirical correlation relationships have been found for a wide variety of geometries applicable to the food industry. An extensive list is available in *Perry's Chemical Engineer's Handbook*. However, as in the case of the convective heat-transfer coefficient empirical correlation relationships, it must be remembered that these correlation relationships can provide only an estimate of the actual value of k_m in any given situation.

A correlation that is frequently used in the food industry is for flow over a flat plate, since many foods present relatively flat surfaces to their environments that are prone to drying out. Since the thickness of the boundary layer containing transferring mass increases as the fluid moves across the plate, k_m is not a constant. However, the average value is what we need in order to find the total mass transfer from the surface. In the case of laminar flow over a flat plate ($N_{Re} < 5 \times 10^5$), the average value of k_m can be estimated from the relationship,

$$N_{Sh} = 0.664 \, (N_{Re})^{1/2} \, (N_{Sc})^{1/3} \tag{11.10}$$

where the characteristic dimension, *D*, is the length of the plate in the direction of flow. Other correlation relationships are available that can provide the local value of k_m. In the case of turbulent flow over a flat plate ($N_{Re} > 5 \times 10^5$), the average value of k_m is found using

$$N_{Sh} = 0.036 \, (N_{Re})^{0.8} \, (N_{Sc})^{1/3}. \tag{11.11}$$

Worked problem 11.5

A 9-in. square dish is filled with water and set squarely in front of a fan blowing air across the surface at 5 m/s. The air is 30°C and 80% RH. The diffusivity of water in air is 0.258×10^{-4} m^2/s; the density of the air is 1.127 kg/m^3; and the viscosity of the air is 18.682×10^{-6} Pa·s. How much water will evaporate in 3 h?

1. *Motivation*: This problem looks at convective moisture transfer from a moist flat surface to air when the air is moving across the surface. This situation occurs when any flat, moist food material is exposed to moving air, such as slabs of meat in a refrigerated cooler, sheet cakes in a convection oven, and trays of thin-cut food in a drier. In some cases we want to limit the moisture loss, while in others we want to encourage it. This type of calculation allows us to estimate the moisture losses.
2. *Define the problem*: We are given the conditions of air moving over the flat surface of the water in the dish, as well as the dimensions of the moist surface, and we are asked to determine how much moisture is lost in a given period of time.

3. *Think about the problem*: The geometry is a flat plate, and the air motion indicates that convective mass transfer will occur from the water surface to the bulk air. Near the water surface, the concentration will be constant at the saturation vapor pressure at 30°C, while the concentration in the bulk air will be 80% of that value according to the definition of the relative humidity.

4. *Plan the solution*: We are not given the convective mass-transfer coefficient, k_m. However, we are given the information to determine the Schmidt number, as well as the Reynolds number, which will allow us to find the flow regime. This is the situation of flow over a flat plate covered by equation (11.10) for laminar flow, and equation (11.11) for turbulent flow, for calculation of the Sherwood number, where the equation chosen will depend on the value of the Reynolds number we calculate. We can then calculate k_m from the Sherwood number. After calculating k_m, we can the get the mass-transfer rate from equation (11.8) using the partial pressure of water vapor at the surface and in the bulk air. Then the mass transferred in 3 h is found by multiplying the rate times the time.

5. *Implement the solution strategy*: First we calculate the Reynolds number to determine the flow regime,

$$N_{Re} = \frac{vD}{\mu} = \frac{1.127 \text{ kg/m}^3 \cdot 5 \text{ m/s} \cdot 9 \text{ in.} \cdot 0.0254 \text{ m/in.}}{18.682 \times 10^{-6} \text{ Pa·s}} = 68,952$$

which is laminar. Next we calculate the Schmidt number,

$$N_{Sc} = \frac{\rho}{\mu D_{AB}} = \frac{18.682 \times 10^{-6} \text{ Pa·s}}{1.127 \text{ kg/m}^3 \cdot 0.258 \times 10^{-4} \text{ m}^2/\text{s}} = 0.643.$$

Then we use the N_{Re} and N_{Sc} to calculate the Sherwood number, and therefore the convective mass-transfer coefficient, for laminar flow over a flat plate using equation (11.10)

$$N_{Sh} = \frac{k_m D}{D_{AB}} = 0664(68952)^{1/2}(0.643)^{1/3} = 150.45$$

$$k_m = N_{Sh} \frac{D_{AB}}{D} = 150.45 \frac{0.258 \times 10^{-4} \text{ m}^2/\text{s}}{0.2286 \text{ m}} = 0.017 \text{ m/s}.$$

From the steam table at 30°C,

$$p^{\circ}_{w} = 4.246 \text{ kPa}.$$

Based on the definition of the relative humidity,

$$p_{w\infty} = 80/100 \cdot p^{\circ}_{w} = 3.397 \text{ kPa}.$$

Then we can use equation (11.8) to calculate the mass-transfer rate,

$$m_w = \frac{k_m M_w A}{RT}\left(p_w^o - p_{w\infty}\right) = \frac{0.017 \ \sqrt[m]{s} \cdot 18 \ \sqrt[kg]{kg\,mol} \cdot (0.2286 \ m)^2}{8.314 \ \sqrt[m^3 \cdot kPa]{(kg\,mol \cdot K)} \cdot (30 + 273.15)K}$$

$$\times \ (4.246 \ kPa - 3.397 \ kPa) = 5.328 \times 10^{-6} \ kg/s.$$

The final answer is then found by multiplying the mass-transfer rate by the time,

$$m_w = 5.328 \times 10^{-6} \ kg/s \cdot 1000 \ g/kg \cdot 3600 \ s/h \cdot 3 \ h$$

$$m_w = 58.1 \ grams \ of \ water \ lost \ from \ the \ dish \ in \ 3 \ h.$$

6. *Check the solution:* No independent verification of this calculation is possible; however, we would expect a rate of moisture loss of this magnitude from an open dish of water.
7. *Generalize the solution strategy:* This technique can be used to calculate the mass transfer of any free liquid from a solid surface to air, as long as mass transfer through a solid matrix to the surface is not involved.

Practice problem 11.6

A 1-mm droplet of liquid milk is injected into 80°C air in a spray drier with a relative velocity between the air and the droplet of 1 m/s. The temperature of the droplet is 30°C, which is also the wet-bulb temperature of the air. At 80°C, the diffusivity of water in air is 3.33×10^{-5} m^2/s; the density of the air is 0.968 kg/m^3; and the viscosity of the air is 20.79×10^{-6} Pa·s. The Sherwood number for mass transfer from spheres in laminar flow with the sphere diameter as the characteristic dimension is

$$N_{Sh} = 2 + 0.6 \ (N_{Re})^{1/2} \ (N_{Sc})^{1/3}. \tag{11.12}$$

What is the initial transfer rate of water from the droplet?

(Hint: the surface area of a sphere is πd^2.)

11.4   *Unsteady-state mass transfer*

As mentioned in the introduction, there are many instances in food processing where the rate of mass transfer changes over time. This is typically due to the reduction of the chemical-potential gradient of the concentration difference driving the mass transfer as it occurs. The relationship used to describe unsteady-state mass transfer into an infinite slab that has a half thickness of x is

$$\frac{\partial C_A}{\partial t} = D_{AB} \frac{\partial^2 C_A}{\partial x^2} \tag{11.13}$$

assuming constant physical properties. This relationship is in the exact same form as the equation for unsteady-state heat transfer (equation [10.31]). Therefore, we can make use of the graphical solutions for unsteady-state heat transfer such as the Heisler or Gurney–Lurie charts for infinite slab, infinite cylinder, and sphere standard geometries by replacing the heat-transfer dimensionless numbers with analogous mass-transfer dimensionless numbers. This includes replacing the Biot number in equation (10.32), the unaccomplished temperature difference in equation (10.33), and the Fourier number in equation (10.34) with the mass-transfer Biot number:

$$N_{Bi,m} = \frac{k_m D}{D_{AB}} \tag{11.14}$$

the concentration ratio:

$$\mathcal{P} = \frac{C_A - C_{AM}}{C_{Ai} - C_{AM}} \tag{11.15}$$

and the mass-transfer Fourier number:

$$N_{Fo,m} = D_{AB} \frac{t}{D^2} \tag{11.16}$$

where k_m is the convective mass-transfer coefficient for the mass transfer of A from the bulk fluid medium (M) to the surface of B; D_{AB} is the diffusivity of A in B; C_A is the concentration of A in B at the characteristic dimension (D = half thickness of an infinite slab, radius of infinite cylinder, and radius of a sphere); C_{Ai} is the initial concentration of A in B; C_{AM} is the concentration of A in the bulk medium; and t is the time that has passed. There are also unsteady-state mass-transfer graphical solutions for the mass average concentration ($\bar{C}_A$) in solids for the three standard geometries, i.e., Treybal chart (Treybal, 1980). These charts make use of the concentration ratio with C_A replaced with $\bar{C}_A$. They are applicable to situations where there is negligible resistance to mass transfer at the surface, which is generally the case for mass transfer from a liquid or a gas into a solid.

Worked problem 11.7

Individual pieces of dry spaghetti, which can be considered as infinite cylinders with a diameter of 2 mm, are exposed to an environment at 25°C and 80% RH. The mass diffusivity for water vapor within the pasta is 1.25×10^{-11} m²/s. If the initial water activity of the pasta is 0.05, estimate the mean water activity of the pasta piece after being held at these conditions for half a day. (Hint: the relationship of water activity to partial pressure of water vapor in air is given in equation [6.21].)

1. *Motivation*: This problem looks at moisture transfer into an unprotected food material. Determination of the change in the water activity due to the uptake of moisture when exposed to the environment for a specified amount of time will allow us to determine the effect on the shelf life of dry food materials.

2. *Define the problem*: We are given the average conditions of the outside air and the initial moisture condition of the food material, as well the mass diffusivity for water in the food material, and are asked to determine the average change in the moisture condition of the food material after it has been exposed to the environment for a day.

3. *Think about the problem*: The relationship of the water activity to the water vapor pressure of air from chapter 6, along with the definition of the relative humidity, allows us to relate the concentration ratio to the water activity in terms of the vapor pressure. Spaghetti is extremely long in comparison to its radius, so end effects will be negligible, and it can be considered as an infinite cylinder. Since the main resistance to mass transfer of the water vapor is within the solid pasta cylinder, all other resistances are neglected, and we can use the Treybal chart curve for an infinite cylinder to find the concentration ratio.

4. *Plan the solution*: The mass-transfer Fourier number is calculated using the time converted to seconds, the diffusivity of water in pasta, and the spaghetti radius. The mass average concentration ratio is then found on the mass average concentration, unsteady-state mass-transfer (Treybal) chart using the mass-transfer Fourier number and the curve for the infinite cylinder. The relative humidity is used to determine the water activity of the medium, and the initial water activity is given. Then the mass average water activity can be calculated from the mass average concentration ratio.

5. *Implement the solution strategy*: Using equation (11.16),

$$N_{Fo,m} = D_{AB} \frac{t}{D^2}$$

$$= 1.25 \times 10^{-11} \text{m}^2/\text{s} \frac{0.5 \text{ day} \cdot 24 \text{ h/day} \cdot 60 \text{ min/h} \cdot 60 \text{ s/min}}{(0.001 \text{ m})^2}$$

$$= 0.54.$$

From a Treybal chart using the curve for an infinite cylinder,

$$\bar{Y}_m = 0.03.$$

Using equation (6.21) and the definition of the relative humidity, we can convert the mass average concentration ratio version of equation

(11.15) to be in terms of the water activity:

$$a_w = \frac{p_w}{p_w} = \frac{\%\mathrm{RH}}{100} \rightarrow \bar{Y}_m = \frac{\bar{C}_w - C_{wM}}{C_{wi} - C_{wM}} = \frac{\bar{p}_w - p_{wM}}{p_{wi} - p_{wM}} = \frac{\bar{a}_w - \frac{\%\mathrm{RH}}{100}}{a_{wi} - \frac{\%\mathrm{RH}}{100}}.$$

Solving for the mean water activity:

$$\bar{a}_w = \bar{Y}_m \cdot \left(a_{wi} - \frac{\%\mathrm{RH}}{100} \right) + \frac{\%\mathrm{RH}}{100}$$

$$\bar{a}_w = 0.03 \cdot (0.05 - 0.8) + 0.8 = 0.78$$

6. *Check the solution*: According to this calculation, a very thin piece of spaghetti will have nearly come to equilibrium with its environment after a day. No independent verification of this calculation is possible; however, we would expect that a very thin piece of material like this would quickly come to equilibrium with its environment.
7. *Generalize the solution strategy*: This approach can be used to follow the change in mass average concentration of a component in a solid with a standard shape over time to or from a gas, vapor, or liquid with an initial concentration-based chemical potential difference.

Practice problem 11.8

Determine the water activity in the center of the piece of spaghetti in worked problem 11.7, using a mass-transfer coefficient for the environment around the pasta of $k_m = 1.2 \times 10^{-4}$ m/s. (Note: solution of this problem requires you to use a graphical solution for unsteady-state heat transfer from another source.)

References

Perry, R. H., and Green, D. 1997. *Perry's Chemical Engineers' Handbook*, 7th ed. New York: McGraw-Hill Professional.

Singh, R. P., and Heldman, D. R. 2001. *Introduction to Food Engineering*, 3rd ed., San Diego: Academic Press.

Treybal, R. E. 1980. *Mass Transfer Operations*, 3rd ed. New York: McGraw-Hill.

Appendix 1: Common conversion factors for engineering units

Length	Power
1 in. = 2.54 cm	1 hp = 0.7457 kW
100 cm = 1 m	1 W = 14.340 cal/min
1 micron = 10^{-6} m = 10^{-4} cm = 10^{-3} mm = 1 μm	1 hp = 550 ft lb_f/s
1 mile = 5280 ft	1 Btu/h = 0.29307 W
1 ft = 12 in.	1 hp = 0.7068 Btu/s
1 m = 3.2808 ft = 39.37 in.	1 J/s = 1 W

Volume	Mass
1 L = 1000 cm^3	1 lb_m = 453.59 g = 0.45359 kg
1 m^3 = 1000 L	1 kg = 2.2046 lb_m
1 in^3 = 16.387 cm^3	1 ton = 2000 lb_m
1 US gallon = 3.785 L	1 ton (metric) = 1000 kg

Force	Density
1 g cm/s^2 (dyn) = 10^{-5} kg m/s^2 = 10^{-5} N	1 g/cm^3 = 62.43 lb_m/ft^3 = 1000 kg/m^3
1 lb_f = 4.4482 N	1 lb_m/ft^3 = 16.0185 kg/m^3

Pressure	Acceleration Due to Gravity
1 bar = 10^5 Pa = 10^5 N/m^2	g = 9.807 m/s^2
1 psia = 1 lb_f/in^2 = 6.895×10^3 Pa	g = 32.174 ft/s^2
1 atm = 14.696 psia = 1.01325×10^5 Pa = 1.0325 bar	g_c = 32.174 ($lb_m \cdot$ft)/(s^2 lb_f)
1 atm = 29.92 in. Hg = 760 mm Hg @ 0°C	g_c = 1 (kg·m)/($s^2 \cdot$N)

Thermal Conductivity	Heat-Transfer Coefficient
1 Btu/(h·ft·°F) = 1.7307 W/(m·K)	1 Btu/(h·$ft^2 \cdot$°F) = 5.6783 W/($m^2 \cdot$K)

Heat, Energy, Work	Gas Law Constant (R)
$1\ J = 1\ N{\cdot}m = 1\ (kg{\cdot}m^2)/s^2$	$1.9872\ (g{\cdot}cal)/(g\ mol{\cdot}K)$
$1\ Btu = 1055.06\ J$	$1.9872\ Btu/(lb\ mol{\cdot}K)$
$1\ Btu = 252.16\ cal$	$82.057\ (cm^3{\cdot}atm)/(g\ mol{\cdot}K)$
$1\ kcal = 1000\ cal = 4.184\ kJ$	$8314.34\ J/(kg\ mol{\cdot}K)$
$1\ Btu = 778.17\ ft{\cdot}lb_f$	$8314.34\ (kg{\cdot}m^2)/(s^2{\cdot}kg\ mol{\cdot}K)$

Viscosity	Temperature
$1\ cP = 10^{-2}\ g/(cm{\cdot}s) = 10^{-2}\ P$	$K = 273.15 + {}^{\circ}C$
$1\ cP = 10^{-3}\ Pa\cdot s$	${}^{\circ}F = {}^{\circ}C(1.8) + 32$
	${}^{\circ}C = ({}^{\circ}F - 32)/1.8$
	$R = 459 + {}^{\circ}F$
	Note that $\Delta{}^{\circ}F = \Delta{}^{\circ}C(1.8)$

Appendix 2: Answers to practice problems

Chapter 6: Gases and Vapors

6.1: Pressure
Problem 6.2: $P_{vac} = 759.95$ mm Hg
6.2: Gas Laws
Problem 6.4: $x_{O_2} = 0.105$
6.3: Gas Mixtures

6.3.2: Partial Volume
Problem 6.6: $P = 42.26$ kPa

Chapter 7: Mass Balances

7.1: Steady-State Operation, No Reaction
Problem 7.2: $P = 50$ lb/h
Problem 7.4: $P = 230$ kg with 73.5% solids
Problem 7.6: $x_a = 0.159$ alcohol, $x_s = 0.00775$ sugar
Problem 7.8: $x_{H2O} = 0.246$; $x_{CSS} = 0.361$; $x_S = 0.393$ with 75.4% total solids
7.2: Steady-State Operation, with Reaction
Problem 7.10: 2% CH_4, 38% CO_2, 36% H_2O, 24% O_2
7.3: Unsteady-State Operation, No Reaction
Problem 7.12: $t = 17.5$ h

Chapter 8: Energy Balances

8.1: Steam Tables 8.1.1: Saturated Steam
Problem 8.2: $P = 3.767$ psia
Problem 8.4:
1. $T = 201.8°F$
2. $q = 659$ Btu
3. $q = 4883.6$ Btu
8.1.2: Superheated Steam
Problem 8.6: $\Delta H = 35.1$ Btu/lb
8.2: Enthalpy Balances
Problem 8.8:
1. $\Delta T = 0.28°C$
2. $\Delta T = 2.83°C$

Chapter 9: Fluid Mechanics

9.1. Rheology
9.1.2: Measurement of Rheological Properties
       Problem 9.2: $n = 0.475$, $k = 90.43$ Pa·s^n
9.2: Fluid Flow
9.2.1: Continuity Equation
       Problem 9.4: $d_i = 3.5$ in.
9.2.2: Determination of Flow Regime
       Problem 9.6:
         1. $N_{Re} = 16{,}914$; yes, flow is turbulent
         2. $N_{Re} = 171.6$; no, flow is laminar
9.2.3: Flow of a Newtonian Fluid in a Pipe
       Problem 9.8:
         1. $\Delta P = 40.5$ Pa
         2. $\Delta P = 247.2$ Pa
9.2.4: Effects of Friction on Fluid Flow
       Problem 9.10:
         1. $N_{Re} = 894$, $f = 0.018$, $L = 27.6$ m
         2. $N_{Re} = 13{,}894$, $f = 0.0073$, $L = 0.075$ m
9.2.5: Mechanical Energy Balance Equation in Fluid Flow
       Problem 9.12: $E_p = 231$ m^2/s^2 = 231 J/kg
9.3. Non-Newtonian Fluid Flow
       Problem 9.13: $N_{GRe} = 30.4$, $\Delta P/L = 30.4$ kPa
       Problem 9.14: $\mu_{eff} = 0.525$ Pa·s

Chapter 10: Heat Transfer

10.1: Steady-State Heat Transfer
10.1.1: Conduction Heat Transfer
       Problem 10.2: $T = 92.5$°C
       Problem 10.4: $q = 776{,}449$ W
       Problem 10.5: $q_{wall} = 713.2$ W, $q_{window\ and\ wall} = 7913$ W; there is more than a tenfold increase
10.1.2: Convection Heat Transfer
       Problem 10.7: $q/A = 5.74$ Btu/h·ft^2
       Problem 10.9: $q/A = 14.54$ W/m, a little lower
10.1.3: Convective Heat-Transfer Coefficients
       Problem 10.11: $h = 7.208$ W/m^2·K
       Problem 10.13: $h = 1{,}032.6$ W/m^2·K; water provides much better heat transfer than air
10.1.4: Heat Exchangers
       Problem 10.15: $L = 10.2$ m; this is slightly longer than the length of the countercurrent heat exchanger
10.2: Unsteady-State Heat Transfer
       Problem 10.17: $h = 207.7$ W/m^2·K

10.3: Freezing

> **Problem 10.19**: $t_f = 28,425$ s = 7.9 h; the apples will not freeze within the 2-h time limit

10.4: Radiation Heat Transfer

> **Problem 10.21**: $q/A = 102.9$ W/m²

Chapter 11: Mass Transfer

11.1: Psychrometrics

11.1.3: Using the Psychrometric Chart To Find the State of Air during Processing

> **Problem 11.2**: $T_{final} \approx 108°C$

11.2: Molecular Diffusion

> **Problem 11.4**: $m_{H2O}/t = 0.0079$ mg/day

11.3: Convective Mass Transfer

> **Problem 11.6**: $\dot{m}_w = 5 \times 10^{-12}$ kg/s initially

11.4: Unsteady-State Mass Transfer

> **Problem 11.8**: $N_{Bi,m} = 9600$ or $1/N_{Bi,m} \cong 0$, $\theta \cong 0.06$ as read from the Heisler chart for an infinite cylinder, $a_w \cong 0.75$ in the center

Appendix 3a: Quiz

You can use any type of calculator on this quiz.

1. Given that $p_1 = 0.50$, $p_2 = 0.30$, $n_1 = 100$, and $n_2 = 80$, evaluate:

$$(p_1 - p_2) + 1.96\sqrt{\frac{p_1(1-p_1)}{n_1} + \frac{p_2(1-p_2)}{n_2}}$$

2. A plumber needs to compare the capacity of two circular pipes with diameters 2 cm and 3 cm. The capacity is proportional to the circular cross-sectional area of the pipe (Figure A3-Q2).
   a. Calculate the cross-sectional area of each pipe.

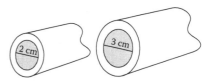

Figure A3-Q2

   b. Express the cross-sectional area of the 2-cm pipe as a percentage of the cross-sectional area of the 3-cm pipe.

3. Does the function $y = e^{3x}$ satisfy the differential equation $\frac{dy}{dx} = 3y$? You must support your answer.

4. The number of bacteria in a culture doubles every 8 hours. The number of bacteria present, A, after t hours is given by the formula

$$A = 500 \times 2^{\frac{t}{8}}$$

a. How many bacteria are present after 48 hours?

b. After how many hours will the culture contain at least 4000 bacteria?

5. In this differential equation, F and V are constants: $0 = Fx + V\frac{dx}{dt}$. Separating variables, it can be written in a differential form as $\frac{dx}{x} = -\frac{F}{V}dt$. Integrate each side of this form of the equation to find an explicit solution that gives x as a function of t.

6. This graph (Figure A3-Q6) shows 10° isotemperature lines over an area. For example, the temperature at position (3, 3) is 50°.

    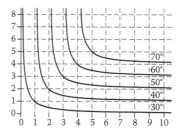

    a. What is the temperature at the position (4, 4)?

Figure A3-Q6

    b. Estimate the temperature at the position (4, 2).

    c. Estimate the temperature at the position (1.5, 3).

    d. Give the coordinates for two different positions with a temperature of 35°.

    e. Draw an arrow at (2, 2) that points in the direction of the most rapid increase of temperature.

7. Commercial airline pilots landing at the Los Angeles International Airport approach the runway along a glide slope that makes a 3° angle with the runway. They can use this table to manually maintain the approach angle. The table relates the aircraft's rate of descent, in feet per minute, to its ground speed, measured in knots.

Ground speed (knots)	70	90	100	120	140	160
Rate of descent (fpm)	379	487	541	649	757	866

Use linear interpolation (Figure A3-Q7) to estimate the following values. (You may wish to plot the values in the table on the grid to help with this problem, although it is not necessary to do so.)

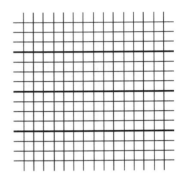

a. If an aircraft is approaching the runway at a ground speed of 110 knots, what rate of descent (in fpm) will maintain the appropriate glide slope?

Figure A3-Q7

b. An aircraft is descending at a rate of 406 fpm. What ground speed should the pilot maintain to approach the runway at the correct angle?

8. A cleaning solution is made up by combining one part phosphoric/ glycolic acid concentrate with eight parts of water. A food technician has 10 liters of the cleaning solution already prepared and needs to increase this to 25 liters.

   a. How many liters of acid concentrate and how many liters of water should be added to the existing 10 liters of cleaning solution to obtain 25 liters at the correct dilution?

   b. The acid concentrate is a 10% solution (i.e., it is 90% water, 10% acid). What is the concentration of the diluted solution?

Appendix 3b: Quiz with answers

1. Given that $p_1 = 0.50$, $p_2 = 0.30$, $n_1 = 100$, and $n_2 = 80$, evaluate:

$$(p_1 - p_2) + 1.96 \sqrt{\frac{p_1(1-p_1)}{n_1} + \frac{p_2(1-p_2)}{n_2}}$$

Answer:

$$(0.50 - 0.30) + 1.96 \sqrt{\frac{0.50(1-0.50)}{100} + \frac{0.30(1-0.30)}{80}}$$

$$= 0.20 + 1.96\sqrt{0.0025 + 0.002625} = 0.340$$

See chapter 1, sections 1.1–1.2 to review math concepts used in Problem 1.

2. A plumber needs to compare the capacity of two circular pipes with diameters 2 cm and 3 cm. The capacity is proportional to the circular cross-sectional area of the pipe.
   a. Calculate the cross-sectional area of each pipe.
   b. Express the cross-sectional area of the 2-cm pipe as a percentage of the cross-sectional area of the 3-cm pipe.

   Answer:

   a. $A = \pi \cdot r^2 = \pi$, $\frac{9}{4}\pi$ cm^2
   b. The ratio is $\dfrac{\pi}{\left(\frac{9}{4}\pi\right)} = \frac{4}{9}$, so the 2-cm pipe has approximately 44% of the capacity of the 3-cm pipe.
   Review basic geometry formulas and see chapter 1, sections 1.1–1.2 to review math concepts used in Problem 2.

3. Does the function $y = e^{3x}$ satisfy the differential equation $\frac{dy}{dx} = 3y$? You must support your answer.

   Answer:
   $\frac{dy}{dx} = 3e^{3x} = 3y$, so $y = e^{3x}$ is a solution of the differential equation

   See chapter 4 to review math concepts used in Problem 3.

4. The number of bacteria in a culture doubles every eight hours. The number of bacteria present, A, after t hours is given by the formula

$$A = 500 \times 2^{\frac{t}{8}}$$

   a. How many bacteria are present after 48 hours?
   b. After how many hours will the culture contain at least 4000 bacteria?

   Answer:

   a. $A = 500 \times 2^{\frac{48}{8}} = 32,000$

   b. Solve $4000 = 500 \times 2^{\frac{t}{8}}$

   $$8 = 2^{\frac{t}{8}} \Rightarrow \log_2 8 = \frac{t}{8} \Rightarrow t = 8 \times 3 = 24 \quad \text{hours}$$

   See chapter 3, section and chapter 3, section 3.3 to review math concepts used in problem 4.

5. In this differential equation, F and V are constants: $0 = Fx + V \frac{dx}{dt}$ Separating variables, it can be written in a differential form as $\frac{dx}{x} = -\frac{F}{V} dt$. Integrate each side of this form of the equation to find an explicit solution that gives x as a function of t.

   Answer:

   $\int \frac{dx}{x} = -\int \frac{F}{V} dt$ gives $\ln x = -\frac{F}{V} t + C$ for any constant C. Exponentiating

   each side gives the explicit solution $x = K \cdot e^{-\frac{F}{V} t}$ for K, a constant.

   See chapter 3, section 3 and chapter 4 to review math concepts used in problem 5.

6. This graph shows 10° isotemperature lines over an area. For example, the temperature at position (3, 3) is 50°.
   a. What is the temperature at the position (4, 4)?
   b. Estimate the temperature at the position (4, 2).
   c. Estimate the temperature at the position (1.5, 3).
   d. Give the coordinates for two different positions with a temperature of 35°.
   e. Draw an arrow at (2, 2) that points in the direction of the most rapid increase of temperature.

   Answer:
   a. The temperature is 60°
   b. The temperature is about 46° (or 47°, using interpolation)
   c. The temperature is about 40° (using interpolation to find the position); answers for (b) and (c) should be within 1 degree

d. A variety of answers are possible; (1.5, 1.5) is one
e. The direction of most rapid increase is perpendicular to the level curves, in this case at a 45° angle pointing to the upper right of the graph.

See chapter 3 sections 3.1 and 3.2 to review math concepts used in problem 6.

7. Commercial airline pilots landing at the Los Angeles International Airport approach the runway along a glide slope that makes a 3° angle with the runway. They can use this table to manually maintain the approach angle. The table relates the aircraft's rate of descent, in feet per minute, to its ground speed, measured in knots.

Ground speed (knots)	70	90	100	120	140	160
Rate of descent (fpm)	379	487	541	649	757	866

Use linear interpolation to estimate the following values. (You may wish to plot the values in the table to help with this problem, though it is not necessary to do so.)

a. If an aircraft is approaching the runway at a ground speed of 110 knots, what rate of descent (in fpm) will maintain the appropriate glide slope?
b. An aircraft is descending at a rate of 406 fpm. What ground speed should the pilot maintain to approach the runway at the correct angle?

Answer:
a. Because 110 is halfway between 100 and 120, the rate of descent should be halfway between 541 and 649: $\frac{541+649}{2} = 595$ fpm
b. $\frac{406-379}{487-379} = \frac{27}{108} = \frac{1}{4}$, so the appropriate groundspeed is one-quarter of the way from 70 to 90 knots: 75 knots is the correct speed

See chapter 2 to review math concepts used in problem 7.

8. A cleaning solution is made up by combining one part phosphoric/glycolic acid concentrate with eight parts of water. A food technician has 10 liters of the cleaning solution already prepared and needs to increase this to 25 liters.
a. How many liters of acid concentrate and how many liters of water should be added to the existing 10 liters of cleaning solution to obtain 25 liters at the correct dilution?
b. The acid concentrate is a 10% solution (i.e., it is 90% water, 10% acid). What is the concentration of the diluted solution?

Answer:

a. Let c be the amount of concentrate that is needed and w the amount of water. Then $c + w = 15$ (the additional mixture required) and $8c = w$ (to obtain the correct dilution). Substituting, $9c = 15$, so $c = 5/3$ liters and $w = 40/3$ liters.

b. The diluted solution is $1/9$ of the 10% concentrate, so the final concentration of acid in the solution is $\frac{1}{9} \times 10\% \approx 1.1\%$.

See chapter 1, sections 1.3–1.4 and chapter 5 to review math concepts used in problem 8.

Index